LEBESGUE INTEGRATION

J. H. WILLIAMSON

DOVER PUBLICATIONS, INC.
Mineola, New York

Bibliographical Note

This Dover edition, first published in 2014, is an unabridged republication of the edition originally published in the "Athena Series: Selected Topics in Mathematics" by Holt, Rinehart and Winston, New York, in 1962.

Library of Congress Cataloging-in-Publication Data

Williamson, J. H. (John Hunter), 1926–
 Lebesgue integration / J.H. Williamson. — Dover edition.
  pages cm
 Originally published: New York : Holt, Rinehart and Winston, 1962.
 Includes index.
 ISBN-13: 978-0-486-78977-4
 ISBN-10: 0-486-78977-2
 1. Lebesgue integral. 2. Measure theory. I. Title.
QA312.W53 2014
515'.43—dc23

              2014011311

Manufactured in the United States by Courier Corporation
78977201 2014
www.doverpublications.com

Preface

This is an introductory text on Lebesgue integration, and may be read by any student who knows a little about real variable theory and elementary calculus. Most of the material on sets and functions required in the book is collected together in Chapter 1. The reader who is not already acquainted with these topics should find there enough for an understanding of the chapters that follow. At one or two points a slight acquaintance with linear algebra (matrices and determinants) is helpful. No knowledge of the Riemann integral is required.

It is widely agreed that the Lebesgue integral should be developed in the general setting of n-dimensional Euclidean space, rather than in the special case $n = 1$. This is done here. Apart from one or two initial complications, the more general approach is no more difficult, and is indeed essential for the formulation of some important theorems. Nothing essential will be lost if the reader takes $n = 2$ throughout the book.

The basic material is developed in detail; in later chapters (in particular, Chapter 6) more is left to the reader. The exercises at the ends of the chapters range from the trivial to the difficult, and include one or two substantial theorems. One exercise is, of course, implicit throughout the book: verify in detail all assertions made in the text and not proved there.

There are many excellent advanced texts on integration theory, to which this book may serve as an introduction—for example, P. R. Halmos, *Measure Theory* or A. C. Zaanen, *An Introduction to the Theory of Integration*.

J. H. W.

Cambridge, England
January, 1962

Contents

[1]

Sets and Functions

1.1. Generalities

We take the usual simple-minded analyst's view of set theory. The terms *set*, collection, aggregate, class, and so forth, are looked on as the same and will not be defined. The elements that make up a set may be called *points*, irrespective of their nature (often they will be points, in the usual sense, in Euclidean space). A collection of sets will be called a *class* rather than a set. We shall tend to use lower-case italic letters for points, italic capitals for sets, and script capitals for classes.

As usual, the property that the element a is in the set A (a belongs to A, A contains a) is denoted by $a \in A$. The negation of this is written $a \notin A$. If $a \in A$ implies $a \in B$ (every element in A is also in B), then A is a *subset* of B, and we write $A \subset B$ or $B \supset A$. If $A \subset B$ and $B \subset A$, then the sets A and B are *equal*; $A = B$. If $A \subset B$ and $A \neq B$, then A is a *proper subset* of B. If A consists of the finite number of points $a, b, \cdots, k$, we write $A = \{a, b, \cdots, k\}$. More generally, if A consists of all points a for which the statement $P(a)$ is true, we write $A = \{a: P(a)\}$.

It is convenient to introduce the *empty set*, denoted by $\emptyset$, which contains no points; it is a subset of every set. Also, all sets under consideration at any time will be subsets of some "large" set, the *universal set*, whole space $\cdots$. This will usually be left to be understood from the context; it will almost always be Euclidean space of the appropriate dimension, or a suitable subset of this.

A *map* or mapping f of A into B, or a *function f* from A to B, is a correspondence which assigns to each $a \in A$ an element $f(a) \in B$; $f(a)$ is the *image* of a under f. The set A is the *domain* (of definition) of f; in order to specify a function completely, its domain should be given, but often we will leave this to be understood from the context; no confusion should result. If E is a subset of A, the set $f(E) = \{f(a): a \in E\}$ is the image of E under f. The set $f(A)$ is the *range* of f. If $F \subset B$, the subset of A defined by $\{a: f(a) \in F\}$ is called the *inverse image* of F under f, and it is denoted by $f^{-1}(F)$. If $f(A) = B$, then f is a map *on to B*. If $f(a) = f(a')$ implies $a = a'$, then f is *one to one* (1-1).

Let A be a set. A *family* in A is a set I (the index set) and a mapping f of I into A. The element $f(i)$ may be written a_i, and the family may be denoted by $(a_i)_{i \in I}$ or simply (a_i). The most familiar example, of course, is where I is

1

the set of integers from 1 to n, or the set of all positive integers; the family is then a finite or infinite sequence. A family in A is thus something more complicated than a subset of A; naturally, to each family (a_i) there corresponds a subset $f(I)$ of A. In the particular case where the map f is 1-1, the family (a_i) contains no repetitions, and is then an indexed set. Any set can be regarded as an indexed set (take the set itself as index set), and hence as a family. The set corresponding to the family (a_i) may be denoted $\{a_i\}$, $\{a_i : i \in I\}$, or $\{a_i\}_{i \in I}$.

If A and B are any sets, their *union* $A \cup B$ is the set of points which are in at least one of A, B, and similarly for any finite class. More generally, if $\mathcal{O}$ is a class or $(A_i)_{i \in I}$ a family of sets, the set of points which belong to at least one set in the class or family is the union of the sets in question, denoted $\bigcup_{A \in \mathcal{O}} A$ or $\bigcup_{i \in I} A_i$. Sometimes we simply write $\bigcup A$ when is no risk of confusion. The *intersection* of A and B, $A \cap B$, is the set of all points which are in both A and B; this extends to $\bigcap_{A \in \mathcal{O}} A$, $\bigcap_{i \in I} A_i$ in the obvious way. If $A \cap B = \varnothing$, A and B are said to be *disjoint*; more generally, a class $\mathcal{O}$ of sets is called disjoint if each pair of sets in $\mathcal{O}$ is disjoint.

Unions and intersections have various algebraic properties; for instance, the associative properties $A \cup (B \cup C) = (A \cup B) \cup C = A \cup B \cup C$ and $A \cap (B \cap C) = (A \cap B) \cap C = A \cap B \cap C$; the obvious commutative relations, and the two distributive laws $(A \cup B) \cap C = (A \cap C) \cup (B \cap C)$ and $(A \cap B) \cup C = (A \cup C) \cap (B \cup C)$. The proof in each case is immediate, and, of course, very substantial generalizations are equally easy. One consequence of the identities

$$\bigcup_{A \in \mathcal{O}} A \cup \bigcup_{A \in \mathcal{B}} A = \bigcup_{A \in \mathcal{O} \cup \mathcal{B}} A \quad : \quad \bigcap_{A \in \mathcal{O}} A \cap \bigcap_{A \in \mathcal{B}} A = \bigcap_{A \in \mathcal{O} \cup \mathcal{B}} A$$

is worth noting; taking $\mathcal{B} = \varnothing$, it is seen that we must have

$$\bigcup_{A \in \theta} A = \varnothing, \qquad \bigcap_{A \in \theta} A = X,$$

where X is the universal set. A rather similar situation arises with empty sets of real numbers; we must have

$$\sup_{r \in \theta} r = -\infty, \qquad \inf_{r \in \theta} r = +\infty.$$

The *difference* $A \setminus B$ is the set of points of A which are not in B; B need not be a subset of A. If B is a subset of A, then $A \setminus B$ is also called the *complement* of B with respect to A, and is denoted $\mathcal{C}_A B$. If A is the universal set X, we speak simply of the complement of B, and write $\mathcal{C}B$. For any A, B, we have $A \setminus B = A \cap \mathcal{C}B$. It is clear that $A \subset B$ is equivalent to $\mathcal{C}A \supset \mathcal{C}B$, that $\mathcal{C}(\bigcup A) = \bigcap \mathcal{C}A$, and that $\mathcal{C}(\bigcap A) = \bigcup \mathcal{C}A$. From this there follows the very useful *principle of duality*: from any set-theoretic relation, we can obtain another by reversing inclusions, replacing unions by intersections and intersections by unions, and replacing each set by its complement. In particular, from any formal identity—such as $(A \cup B) \cap C = (A \cap C) \cup (B \cap C)$—we can obtain another, equally true, by interchanging unions and intersections; in the present case the result is $(A \cap B) \cup C = (A \cup C) \cap (B \cup C)$, the other distributive law. The principle can be extended to cover topological situations; the complement of an open set is closed and the complement of a closed set is open (see Theorem 1.3d).

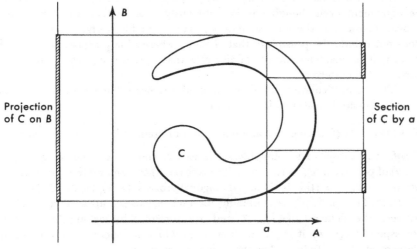

Fig. 1

The *Cartesian product* $A \times B$ of A and B is the set of all pairs (a, b) with $a \in A$, $b \in B$; this is true also for any finite number of factors. The product is not commutative; it is, however, associative: $(A \times B) \times C = A \times (B \times C) = A \times B \times C$ in the sense that there is a natural correspondence between $((a, b), c)$, $(a, (b, c))$ and (a, b, c). The product $R \times R$ of the real line with itself is the Euclidean plane R^2 (hence the name Cartesian, after Descartes) and the product of n factors R is n-dimensional Euclidean space R^n. The relation $R^m \times R^n = R^{m+n}$ holds in the sense explained above. The notion may be generalized to situations like $\underset{A \in \alpha}{\Large\times} A$ or $\underset{i \in I}{\Large\times} A_i$. A very convenient pictorial representation of the Cartesian product of two sets is obtained by taking each to be a segment of the real line, in which case their product is a subset of the plane.

Let C be a subset of $A \times B$, and $a \in A$. By the *section* of C by a we mean the set $\{b: (a, b) \in C\}$; it is a subset of B which may, of course, be empty. Again, the notion can be generalized to products $\underset{i \in I}{\Large\times} A_i$. The *projection* of C on B is the set $\{b: (a, b) \in C$ for some $a \in A\}$. It is the union of all the sections of C by points of A (Fig. 1).

1.2. Countable and Uncountable Sets

The sets A and B are *equivalent* (or *similar*) if there is a 1-1 map of one on to the other; we write $A \sim B$. It is clear that $A \sim A$, that $A \sim B$ if and only if $B \sim A$, and that $A \sim B, B \sim C$ imply $A \sim C$. If $A = \emptyset$, or $A \sim \{1, 2, \cdots, n\}$ for some n, then A is *finite*; if A is finite, or equivalent to the set of all positive integers, it is *countable*.

Infinite sets have some surprising properties; for instance, two sets may be equivalent even though one is "obviously" larger than the other. (It is proved below that the rational numbers are countable.) The relation $A \sim B$ does not exlude the possibility that $A \sim C$, where C is a proper subset of B; indeed, it is characteristic of infinite sets that they are equivalent to proper subsets of themselves.

The term "countable union" is used to mean "union of a countable class"; this applies also to intersection.

THEOREM 1.2a (Cantor). *A countable union of countable sets is countable.*

Proof. Any subset of a countable set is clearly countable. The set S of pairs (m, n) of positive integers is countable; make (m, n) correspond to $\frac{1}{2}(m + n - 2)$ $(m + n - 1) + m$ (this is the usual diagonal process: $(1, 1)$, $(1, 2)$, $(2, 1)$, $(1, 3)$, $(2, 2)$, $(3, 1)$, $\cdots$). If A_n is countable for each n, we set up a correspondence between the elements of $\bigcup_{n=1}^{\infty} A_n$ and a subset of S by making the element a correspond to (m, n) if A_m is the first set in which a appears, and a is the nth element of A_m. It follows that the union is countable.

COROLLARY. *The set R_0 of rational numbers is countable.*

Proof. The set A_n of rational numbers p/q with $|p| + q \leq n$ is finite for each n, and $R_0 = \bigcup_{n=1}^{\infty} A_n$.

THEOREM 1.2b. *The real numbers are not countable.*

Proof. It is enough to consider the real numbers r such that $0 < r \leq 1$; suppose that these are enumerated as $r_1, r_2, r_3, \cdots$. For each m, let r_m be represented as a nonterminating decimal:

$$r_m = 0 \cdot x_{m1} x_{m2} x_{m3} \cdots$$

where each x_{mn} is an integer between 0 and 9 (inclusive). Now let $x_n = x_{nn} + 1$ if $x_{nn} < 5$, $x_n = x_{nn} - 1$ if $x_{nn} \geq 5$. Then $r = 0 \cdot x_1 x_2 x_3 \cdots$ is evidently a real number with $0 < r \leq 1$, and $r \neq r_m$ for all m—a contradiction.

It is very easy to see (either directly, or as a corollary of Bernstein's theorem, which follows) that the open unit interval $]0, 1[= \{x: 0 < x < 1\}$, the closed interval $[0, 1] = \{x: 0 \leq x \leq 1\}$, and the two half-open unit intervals are all equivalent. This being so, we can define a *continuum* to be any set equivalent to the unit interval $(0, 1)$.

The next theorem not only is of very considerable theoretical interest but also is useful in enabling one to avoid tedious and irrelevant details in discussing examples.

THEOREM 1.2c (Bernstein). *If $A \sim B' \subset B$ and $B \sim A' \subset A$, then $A \sim B$.*

Proof. Let the 1-1 map of A on B' be ϕ, and let the map of B on A' be ψ. Define

$$A_0 = A \setminus A', \qquad\qquad B_0 = B \setminus B',$$

$$A_r = \psi B_{r-1}, \qquad\qquad B_r = \phi A_{r-1} \ (r \geq 1),$$

$$A_\infty = A \setminus \bigcup_{r=0}^{\infty} A_r, \qquad\qquad B_\infty = B \setminus \bigcup_{r=0}^{\infty} B_r.$$

Then it is immediate that $\phi A_\infty = B_\infty$. Now we can define a 1-1 correspondence between A and B by making $A_0 \cup A_2 \cup A_4 \cup \cdots$ correspond to $B_1 \cup B_3 \cup B_5 \cup \cdots$ by ϕ, $A_1 \cup A_3 \cup A_5 \cup \cdots$ correspond to $B_0 \cup B_2 \cup B_4 \cdots \cup$ by ψ^{-1}, and A_∞ correspond to B_∞ by ϕ.

An equivalent formulation of the theorem is: if $A \supset B \supset C$ and $A \sim C$, then $A \sim B$.

The unit square S in R^2 is a continuum. It is clear, on the one hand, that the unit interval I is equivalent to a subset of S. On the other hand, let $(x, y) \in S$ be written as $(0 \cdot x_1 x_2 x_3 \cdots, 0 \cdot y_1 y_2 y_3 \cdots)$, where the nonterminating decimal is taken in ambiguous cases. A map of S on to a (proper) subset of I is then defined by making (x, y) correspond to the point $0 \cdot x_1 y_1 x_2 y_2 \cdots$ of I. Hence, by the above theorem, $S \sim I$. A similar argument proves that the unit cube in R^n is a continuum, for any n.

Also, the real line R itself is a continuum; take the correspondence between R and I defined by $y = \tan(\pi x - \pi/2)$. Similarly, R^n is a continuum for any n; and, indeed, the result extends also to infinite-dimensional spaces. There are sets, however, which are not countable and not equivalent to I, as is shown in Theorem 1.2d below.

The class $\mathscr{S}$ of all sets of positive integers is also a continuum. Any point in I can be represented in the form $2^{-n_1} + 2^{-n_2} + 2^{-n_3} + \cdots$, with the usual ambiguities in certain cases. Taking the infinite set of integers $\{n_1, n_2, \cdots\}$ in such cases, we have a 1-1 map of I on to a subset of $\mathscr{S}$. Also, to each set in $\mathscr{S}$ there corresponds the point $3^{-n_1} + 3^{-n_2} + 3^{-n_3} + \cdots$ of I; no point can arise in this way from two distinct sets in $\mathscr{S}$. Thus $\mathscr{S}$ is mapped on to a subset of I, and the required result follows.

THEOREM 1.2d (Cantor). *Given any set A, there is a set B such that $A \sim B' \subset B$ but $A \nsim B$.*

Proof. Let B be the class of all subsets of A; there is an obvious correspondence between A and the single-element sets $\{a\}$ in B. Suppose $A \sim B$; let S_a be the subset of A corresponding to the element a. Consider the set

$$T = \{a : a \notin S_a\};$$

that is, the subset of A consisting of those elements which do not belong to the subset to which they correspond. By assumption, $T = S_x$ for some $x \in A$. If $x \in T$, that is, $x \in S_x$, we have an immediate contradiction; if $x \notin T$ then $x \notin S_x$ and so $x \in T$, again a contradiction. It follows that A and B are not equivalent.

If A is a countable infinite set, the set B exhibited in the above theorem is a continuum. It is conjectured that there are no sets intermediate (in the sense of equivalence) between a countable infinite set and a continuum; this is the "continuum hypothesis" (in its simplest form).

There is a set, whose construction is the work of Cantor, which is frequently of use in the manufacture of examples. It is a subset of the closed unit interval which is obtained by removing successively the open intervals $]1/3, 2/3[$; $]1/9, 2/9[$ and $]7/9, 8/9[$; $]1/27, 2/27[$, $]7/27, 8/27[$, $]19/27, 20/27[$ and $]25/27, 26/27[$; $\cdots$. More formally, if

$$E_n = \overset{3^{n-1}}{\underset{r=1}{\bigcup}} \,](3r-2)\,3^{-n}, (3r-1)\,3^{-n}[, \qquad E = \overset{\infty}{\underset{n=1}{\bigcup}} E_n,$$

then Cantor's set T is the complement of E with respect to $[0, 1]$. It is easy to characterize the points of T; each point of $[0, 1]$ can be represented (with the usual ambiguities) in the form $\sum_{n=1}^{\infty} x_n 3^{-n}$, where each x_n is 0, 1, or 2. The points of E_n are precisely those for which x_n is necessarily 1, and so the points of T are those which admit a representation in which each x_n is 0 or 2. Thus $2/3 \in T$, since it has a representation with $x_1 = 2$, $x_2 = x_3 = \cdots = 0$, even though it also has a representation with $x_n = 1$ for all n.

Cantor's set is a continuum. Suppose each point of $[0, 1]$ is represented in the form $\sum_{n=1}^{\infty} x_n 2^{-n}$, the nonterminating representation being preferred in ambiguous cases; if $y_n = 0$ when $x_n = 0$, and $y_n = 2$ when $x_n = 1$, the correspondence $\sum x_n 2^{-n} \rightarrow \sum y_n 3^{-n}$ maps $[0, 1]$ on to a (proper) subset of T. The required result follows, by Bernstein's theorem. On the other hand, it is easy to see that T can contain no nondegenerate intervals; any such interval would contain a point of E_n for some n.

It is convenient to describe, at this point, three functions which are associated with T, which will be referred to later. Let f be the map of $[0, 1]$ into T defined in the previous paragraph. Let g be the function (Cantor's function) on $[0, 1]$ to itself defined by

$$g(x) = \sup_{f(y) \leq x} y.$$

It is immediate that g is an increasing function, taking the values 1/2 in [1/3, 2/3], 1/4 in [1/9, 2/9], 3/4 in [7/9, 8/9], $\cdots$. Finally, let h be the function defined by

$$h(x) = y \quad \text{if } x = f(y)$$
$$= 0 \quad \text{if } x \text{ is not of the form } f(y).$$

The function h is not increasing, but it clearly has the property that if $x < x'$ then either $h(x') = 0$ or $h(x) < h(x')$. It follows that if I is any subinterval of [0, 1], then $h(I)$ is the union of an interval and the set $\{0\}$.

The remarkable properties of Cantor's set T are not, of course, essentially connected with the number three; any set of similar construction would enjoy the same properties.

1.3. Sets in R^n

Points ($=$ vectors) in R^n, real n-dimensional Euclidean space, are written as $x = (x_1, \cdots, x_n)$. If c is a real number, the product cx is $(cx_1, \cdots, cx_n)$; the sum $x + y$ is $(x_1 + y_1, \cdots, x_n + y_n)$. If A and B are subsets of R^n, the set $A + B$ is defined to be $\{x + y: x \in A, y \in B\}$; and if C is a set of real numbers CA is $\{cx: c \in C, x \in A\}$.

The (Euclidean, quadratic) *distance* from x to y is

$$d(x, y) = \left(\sum_{r=1}^{n} (x_r - y_r)^2\right)^{1/2}.$$

The (quadratic) *norm* $\| x \|$ of x is $d(x, 0)$, where, as usual, $0 = (0, \cdots, 0)$. The *diameter* of a set A is

$$d(A) = \sup_{x, y \in A} d(x, y);$$

in one dimension the diameter of an interval is equal to its length, and also to its measure (as defined in Chap. 2).

In what follows the sign $\prec$ is used to mean either $<$ or $\leq$. If $\rho > 0$, the set $\{y: d(x, y) \prec \rho\}$ is the *sphere* with center x and radius ρ; the sphere is *open* if the sign is $<$, *closed* if $\leq$. It is convenient to regard the set $\{x\}$ consisting of the single point x as a degenerate closed sphere of zero radius.

It is sometimes useful to consider *extended vectors* $a = (a_1, \cdots, a_n)$, where each a_r may be finite or $\pm \infty$. An *interval* in R^n is the product of n intervals in R; more precisely, if a, b are extended vectors, such that $a_r \leq b_r$, $a_r < \infty$, $b_r > -\infty$ for all r, the interval (a, b) is

$$\{x: a_r \prec x_r \prec b_r, 1 \leq r \leq n\}.$$

If $a_r = b_r$ for one or more integers r, the interval is *degenerate;* in this case the signs associated with r must both be $\leq$. The combinations $-\infty \leq$ and

$\leq \infty$ are excluded. If all signs are $<$, the interval is *open* and may be written $]a, b[$. If all possible signs are $\leq$, the interval is *closed*. If *all* signs are $\leq$ (which can only happen if a, b are actually points of R^n), the interval is written $[a, b]$. If $a, b \in R^n$ (that is, a_r, b_r are finite for all r) the interval (a, b) is *bounded* (or *finite*). Generally, a set is bounded if it is a subset of some bounded interval. The *center* of a bounded interval (a, b) is the point x with $x_r = (a_r + b_r)/2$.

LEMMA 1.3a. If S is an open sphere, center x, there is an open interval containing x and contained in S; if I is an open interval containing x, there is an open sphere, center x, contained in I.

Proof. If the radius of S is ρ, take the open interval with center at x with each side of length $2\rho n^{-1/2}$; if I is $]a, b[$, take the open sphere with center x and radius

$$\rho = \inf_r (b_r - x_r, x_r - a_r).$$

The point x is an *interior point* of the set A if there is an open interval containing x and contained in A. The set of interior points of A is the *interior* of A, int A (or A°); evidently $A \supset$ int A for any set A. The point x is an *adherent point* of A (x adheres to A) if every open interval containing x contains a point of A (which may be x itself). The set of adherent points of A is the *closure* (or adherence) of A, denoted cl A or $\bar{A}$. It is clear that $A \subset$ cl A for all A. If A is $\emptyset$, the empty set, then int $A =$ cl $A = \emptyset$ also; and $R^n =$ int $R^n =$ cl R^n. The relations int $A \subset$ int B, cl $A \subset$ cl B obviously hold whenever $A \subset B$. Finally, if I is any interval, int I is the open interval with the same endpoints and cl I is the closed interval with the same endpoints.

We now define a set to be *open* if it is equal to its interior; that is, if every point of the set is an interior point. It is *closed* if it is equal to its closure; that is, if every adherent point is actually in the set. It is immediate that open intervals and open spheres are open sets according to the definition, and that closed intervals and closed spheres are closed sets. In view of Lemma 1.3a, the definitions could equally well have been given in terms of open spheres: a set is open if and only if each point of the set is the center of an open sphere contained in the set; it is closed if and only if each point which is the center of arbitrarily small open spheres which contain a point of the set is itself in the set.

In one dimension there is a theorem that an open set is a countable disjoint union of open intervals. This does not generalize to the n-dimensional case, although a similar result, with half-open intervals, does hold.

LEMMA 1.3b. An open set is the union of the open intervals which it contains.

Proof. Let A be the open set, B the union of the open intervals contained in A; trivially, $A \supset B$. On the other hand, for each $x \in A$ there is an open interval $I(x)$ with $x \in I(x) \subset A$. Hence $A = \bigcup_{x \in A} \{x\} \subset \bigcup_{x \in A} I(x) \subset B$, and so $A = B$.

THEOREM 1.3c. *Any union of open sets is open; any finite intersection of open sets is open.*

Proof. The first assertion is obvious. To prove the second, let A_m be open for $1 \le m \le M$; either $\cap A_m$ is empty, in which case it is certainly open, or there is a point $y \in \cap A_m$. If so, for each m there is an open interval I_m with $y \in I_m \subset A_m$. Suppose

$$I_m = \{x: a_r^{(m)} < x_r < b_r^{(m)}, 1 \le r \le n\};$$

write

$$a_r = \sup_m a_r^{(m)} \quad \text{and} \quad b_r = \inf_m b_r^{(m)}.$$

Since the set of indices m is finite, $a_r < y_r < b_r$ for $1 \le r \le n$. Then the interval $]a, b[= \{x: a_r < x_r < b_r, 1 \le r \le n\}$ is open, contains y, and is contained in I_m for all m. It is therefore in $\cap I_m \subset \cap A_m$, which proves that $\cap A_m$ is open.

THEOREM 1.3d. *A set is closed if and only if its complement is open.*

Proof. It is immediate from the definitions that a point adheres to a set A if and only if it is not an interior point of $\mathscr{C}A$. It follows that $\mathscr{C} \operatorname{cl} A = \operatorname{int} \mathscr{C}A$, so that $A = \operatorname{cl} A$, which is the same as $\mathscr{C}A = \mathscr{C} \operatorname{cl} A$, is the same as $\mathscr{C}A = \operatorname{int} \mathscr{C}A$; the required result follows at once.

COROLLARY. *Any intersection of closed sets is closed; any finite union of closed sets is closed.*

Proof. This follows at once, using Theorem 1.3c and the principle of duality.

The extension of the principle of duality embodied in Theorem 1.3d is a very useful one. It enables one to obtain (for example) criteria for compactness (see Sec. 1.4) in terms of closed sets or closed intervals, instead of open sets and intervals.

Let X be a subset of R^n. The subset A of X is *open relative to X* (open in X) if for every $x \in A$ there is an open interval I containing x, with $I \cap X \subset A$. Also, A is *closed relative to X* (closed in X) if all the points of X which adhere to A belong to A. Thus, in R, $]0, 1/2]$ is closed relative to $]0, 1[$; and $[0, 1/2[$ is open relative to $[0, 1]$. Any set is both open and closed relative to itself.

LEMMA 1.3e. *If A is open relative to X, then $A = X \cap B$, where B is open in R^n, and any set of this form is open relative to X.*

Proof. Let $x \in A$; there is an open interval $I(x)$ with $x \in X \cap I(x) \subset A$. Then $A \supset \bigcup_{x \in A} (X \cap I(x)) \supset \bigcup_{x \in A} \{x\} = A$; $B = \bigcup_{x \in A} I(x)$ is open and $A = X \cap B$. The converse is obvious.

1.4. Compactness

Bounded closed intervals in R have some important properties, which can be looked on as consequences of the Heine-Borel theorem. It is convenient to introduce some definitions: A class $\mathcal{O}$ of sets is a *covering* of a set B if each point of B is in a set in $\mathcal{O}$. A subclass of $\mathcal{O}$ is a *refinement* if it is still a covering of B; we speak of a covering as *open* if each set in it is open.

THEOREM 1.4a (Heine-Borel). Every covering of a bounded closed interval in R by open intervals has a finite refinement.

Proof. Let $[a, b]$ be the interval, $\mathcal{I}$ the covering. Let F be the set of points $x \in [a, b]$ such that $[a, x]$ can be covered by a finite subclass of $\mathcal{I}$. Let $c = \sup_{x \in F} x$; it is clear that $c \in [a, b]$. There is an interval of $\mathcal{I}$ which contains c; this interval clearly contains points of F, so that $c \in F$. Also, if $c < b$, the interval contains a point $c' \in [a, b]$ with $c' > c$; since $c' \in F$, this contradicts the definition of c. It follows that $c = b$ and $b \in F$, which is what was wanted.

A set is *compact* if every covering by open intervals has a finite refinement.

THEOREM 1.4b. A bounded closed interval in R^n is compact.

Proof. The result holds for $n = 1$, by the previous theorem; suppose that it holds for dimensions 1 and $n - 1$. Let $[a, b]$ be the n-dimensional interval and let $\mathcal{I}$ be the covering of it by open intervals. Each interval of $\mathcal{I}$ is of the form $I' \times I''$, where I' and I'' are open intervals, respectively $(n - 1)$ dimensional and 1 dimensional. For each $t \in [a_n, b_n]$ let $\mathcal{I}(t)$ be the class of intervals $I = I' \times I''$ such that $t \in I''$. Let $\mathcal{I}'(t)$, $\mathcal{I}''(t)$ be the classes of corresponding intervals I', I'' respectively. Then $\mathcal{I}'(t)$ is a covering by open intervals of the section by the hyperplane $x_n = t$ of the interval $[a, b]$. This is a closed $(n - 1)$-dimensional interval; by hypothesis it is compact. There is thus a finite refinement, $\mathcal{I}_0'(t)$ say, of $\mathcal{I}'(t)$; let $\mathcal{I}_0(t)$, $\mathcal{I}_0''(t)$ be the corresponding finite subclasses of $\mathcal{I}(t)$, $\mathcal{I}''(t)$ respectively.

It is clear that $I''(t)$, the intersection of all $I'' \in \mathcal{I}_0''(t)$, is an open interval containing t. The class of all such intervals is a covering of $[a_n, b_n]$ by open intervals, and so has a finite refinement, consisting of—say—$I''(t_1), \cdots, I''(t_k)$. Then it is clear that the class

$$\bigcup_{j=1}^{k} \mathcal{I}_0(t_j)$$

is a covering of $[a, b]$ by open intervals, which is a refinement of the original covering $\mathcal{I}$.

There are many formulations of compactness which are equivalent to the definition given above. The following variation will be useful presently:

LEMMA 1.4c. *A set is compact if and only if every open covering has a finite refinement.*

Proof. If every covering by open sets has a finite refinement, then every covering by open intervals has a finite refinement and the set is compact.

Conversely, suppose the set is compact, and let $\mathcal{O}$ be a covering by open sets. This induces a covering (in view of Lemma 1.3b) consisting of all open intervals which are contained in some set of $\mathcal{O}$. This has a finite refinement, say $\{I_1, \cdots, I_k\}$; if $I_j \subset A_j$ $(1 \leq j \leq k)$, then $\{A_1, \cdots, A_k\}$ is a finite refinement of $\mathcal{O}$.

THEOREM 1.4d. *A closed subset of a compact set is compact.*

Proof. Let A be compact, B a closed subset of A. Let $\mathcal{E}$ be an open covering of B; then $\mathcal{E}$, together with $\mathcal{C}B$, is an open covering of A. It thus has a finite refinement, consisting of—say—$E_1, \cdots, E_k$, and possibly $\mathcal{C}B$. In any case, $\{E_1, \cdots, E_k\}$ covers B and is a refinement of $\mathcal{E}$.

THEOREM 1.4e. *A set is compact if and only if it is closed and bounded.*

Proof. In view of Theorems 1.4b and 1.4d, a bounded closed set is certainly compact.

Let I_m be the open interval with center at the origin and with side $2m$:

$$I_m = \{x: -m < x_r < m, 1 \leq r \leq n\}.$$

The class $\{I_m: 1 \leq m < \infty\}$ clearly forms an open covering of any set; if the set is compact, there is a finite refinement, and so the set is contained in I_m for some m; that is, the set is bounded.

To show that a compact set is closed, let A be compact and $x \notin A$. For each $y \in A$ there are disjoint open intervals $I(y)$ and $I_y(x)$, containing, respectively, y and x. The set of all such $I(y)$ forms an open covering of A; there is a finite refinement $\{I(y_1), \cdots, I(y_k)\}$. Then the intersection of the corresponding intervals $I_{y_1}(x), \cdots, I_{y_k}(x)$ is an open interval containing x and not containing any point of A, so that A is closed.

1.5. Functions

Functions should be taken to be *real* (mapping R^n into R) unless the contrary is stated or implied. From Chapter 3 onward, it will often be desirable to consider *extended-real* functions, which may take the values $+\infty$, $-\infty$ as well as finite real values. It will sometimes be convenient to consider *complex functions*. A complex function may be regarded as a map into R^2, the real Euclidean plane; a natural generalization is a *vector function*, which is a map into some finite-dimensional space R^m. Generally speaking, all the essential difficulties in integration theory are encountered with real functions; extensions to complex or vector functions are easy. For this reason, such extensions will not be emphasized and will often be omitted.

The definitions and theorems that follow are formulated for real functions; most of them extend at once to complex functions, with no formal change, and to vector functions if the absolute value $| \cdot |$ is replaced by the norm $|| \cdot ||$.

Let D be the domain of definition of the function f, $A \subset D$ and $x \in A$. The function is *continuous in* A *at* x if, given $\epsilon > 0$, there exists $\delta > 0$ such that $| f(x) - f(y) | < \epsilon$ whenever $y \in A$ and $d(x, y) < \delta$. In view of Lemma 1.3a, this last condition is equivalent to: given $\epsilon > 0$, there exists an open interval I containing x, such that $| f(x) - f(y) | < \epsilon$ whenever $y \in A \cap I$. The function is *continuous in* (or *on*) A if it is continuous in A at x for each $x \in A$. It is *uniformly continuous in* (or *on*) A if, given $\epsilon > 0$, there exists $\delta > 0$ such that $| f(x) - f(y) | < \epsilon$ whenever x, $y \in A$ and $d(x, y) < \delta$. The function is *continuous at* x if it is continuous in D at x; note that the assertion that f is continuous in A is less restrictive than the assertion that it is continuous at each point of A.

THEOREM 1.5a. *The function f is continuous in A if and only if $A \cap f^{-1}(B)$ is open relative to A whenever B is open.*

Proof. Suppose that f is continuous in A, and that B is open; let $x \in A, f(x) \in B$. There exists $\epsilon > 0$ such that $| f(x) - b | < \epsilon$ implies $b \in B$. By the assumption of continuity there is an open interval I, containing x, such that $| f(x) - f(y) | < \epsilon$ whenever $y \in A \cap I$. Thus $A \cap f^{-1}(B) \supset A \cap I$, and so $A \cap f^{-1}(B)$ is open relative to A.

If, conversely, $A \cap f^{-1}(B)$ is open relative to A for all open B, let $x \in A$ and take B to be the interval $] f(x) - \epsilon, f(x) + \epsilon[$. Then there is an open interval I, containing x, such that $A \cap I \subset A \cap f^{-1}(B)$. Hence $| f(x) - f(y) | < \epsilon$ whenever $y \in A \cap I$; that is, f is continuous in A at x. Since x was any point of A, f is continuous in A.

THEOREM 1.5b. *If A is compact and f is continuous in A, then f is bounded in A and attains its bounds.*

Proof. Each set of the form $f^{-1}(] - k, k[)$ is open relative to A; hence, by Lemma 1.3e, of the form $A \cap B_k$, where B_k is open. Since the sets $\{ f^{-1}(] - k, k[) \}$ form a covering of A, the same is true of the open sets $\{ B_k \}$; since A is compact there is a finite refinement, and so $A \subset B_{k'}$ for some k'; it follows that $A \subset f^{-1}(] - k', k'[)$. That is, $| f(x) | < k'$ for all $x \in A$; f is bounded on A.

Let M be the upper bound $\sup_{x \in A} f(x)$. Each set $A \cap f^{-1}(] - \infty, M - 1/k[)$ is open relative to A, hence of the form $A \cap C_k$, where C_k is open. If now the bound M is not attained, the sets C_k form an open covering of the compact set A. There is a finite refinement, and so $A \subset C_K$ for some K; that is, $A \subset f^{-1}(] - \infty, M - 1/K[)$. Thus $f(x) < M - 1/K$ for all $x \in A$, and $\sup_{x \in A} f(x) < M$, a contradiction. A similar proof holds for the lower bound.

THEOREM 1.5c. *If A is compact and f is continuous in A, then f is uniformly continuous in A.*

Proof. Given $\epsilon > 0$, $x \in A$, there is an open interval $I(x)$ with center at x such that $|f(x) - f(y)| < \epsilon/2$ whenever $y \in A \cap I(x)$. For any interval I, let I' denote the interval with the same center and with sides half the length of the corresponding sides of I. The intervals $\{I'(x)\}$ form an open covering of A; there is therefore a finite refinement, say $\{I'_1, \cdots, I'_k\}$. If now ρ is half the length of the smallest side which appears in any of $I'_1, \cdots, I'_k$, it is clear that if $d(y, z) < \rho$ and $y, z \in A$ then y and z are both in one of the intervals $I_1, \cdots, I_k$. If this interval has center x, we have

$$|f(y) - f(z)| \leq |f(y) - f(x)| + |f(x) - f(z)| < \epsilon/2 + \epsilon/2 = \epsilon,$$

so that the uniform continuity in A is proved.

We take for granted the elements of convergence theory, as applied to real or complex sequences or series. In particular, we shall use the fact that, if a real sequence is increasing and bounded above, then it is convergent. For real sequences the *upper limit* (lim sup) and *lower limit* (lim inf) are defined by

$$\limsup_{r \to \infty} u_r = \lim_{r \to \infty} \sup_{s \geq r} u_s, \qquad \liminf_{r \to \infty} u_r = \lim_{r \to \infty} \inf_{s \geq r} u_s.$$

The upper and lower limits always exist (as finite real numbers or $\pm \infty$). They are finite and equal if and only if the sequence is convergent.

For sequences of functions, defined on a set A, limits and convergence may be defined in the obvious (pointwise) manner: $\lim_{r \to \infty} f_r = f$ means that $\lim_{r \to \infty} f_r(x) = f(x)$ for all $x \in A$. This holds true also for upper and lower limits, if the functions are real. The sequence $\{f_r\}$ is *uniformly convergent* in A to f if, given $\epsilon > 0$, there exists r_0 such that $|f(x) - f_r(x)| < \epsilon$ whenever $r \geq r_0$, for all $x \in A$. Uniform convergence clearly implies pointwise convergence; the converse is trivially untrue.

For real functions the notions of upper and lower envelope are useful. If $(f_i)_{i \in I}$ is any family of real (or extended-real) functions, the *upper envelope* of the family is the function Φ defined by

$$\Phi(x) = \sup_{i \in I} f_i(x),$$

and the *lower envelope* ϕ is defined by

$$\phi(x) = \inf_{i \in I} f_i(x).$$

It is convenient to use the notation of lattice theory and write $\Phi = \bigvee_{i \in I} f_i$, $\phi = \bigwedge_{i \in I} f_i$. Variations such as $f \vee g$, $\bigvee_{r=1}^{\infty} f_r$ should require no explanation. The relations

$$f \vee g = (f + g + |f - g|)/2, \qquad f \wedge g = (f + g - |f - g|)/2)$$

can be verified at once; they are sometimes useful.

For sequences of functions, the upper and lower limits can be characterized in convenient form by

$$\limsup f_r = \bigwedge_{r=1}^{\infty} \left(\bigvee_{s=r}^{\infty} f_s \right),$$

$$\liminf f_r = \bigvee_{r=1}^{\infty} \left(\bigwedge_{s=r}^{\infty} f_s \right),$$

The above definitions and notations apply also, of course, to families and sequences of real numbers, which may be regarded as functions defined on a single point.

If A is any set, the *characteristic function* χ_A of A is defined by

$$\chi_A(x) = 1 \quad \text{if} \quad x \in A$$
$$= 0 \quad \text{if} \quad x \notin A.$$

The *support* of a function is the closure of the set of points at which it is nonzero. To say that a function, defined throughout R^n, is of compact support is the same as to say that there is a bounded interval outside which the function takes only the value zero.

EXERCISES

1. Prove that $f(\bigcup A) = \bigcup f(A)$, $f(\bigcap A) \subset \bigcap f(A)$, $f^{-1}(\bigcup A) = \bigcup f^{-1}(A)$, $f^{-1}(\bigcap A) = \bigcap f^{-1}(A)$. Produce an example in which $f(\bigcap A) \neq \bigcap f(A)$.

2. Prove $(\bigcup_{A \in \alpha} A) \cap C = \bigcup_{A \in \alpha} (A \cap C)$, and its dual.

3. What is the relation between $f(A \setminus B)$ and $f(A) \setminus f(B)$?

4. If the universal set X is the bounded interval $[a, b]$ of the real line, what should be the interpretations of $\sup_{r \in \phi} r$, $\inf_{r \in \phi} r$, and why? Is the conclusion to be modified if $[a, b]$ is replaced by $]a, b[$?

5. If X is a set of real numbers, how should $\Sigma_{r \in \phi} r$, $\Pi_{r \in \phi} r$ be interpreted? Why?

6. If the real function f is additive for disjoint sets ($f(A \cup B) = f(A) + f(B)$ if $A \cap B = \emptyset$), prove that

$$f(A_1 \cup A_2 \cup \cdots \cup A_n)$$
$$= \Sigma f(A_r) - \Sigma f(A_{rs}) + \Sigma f(A_{rst}) - \cdots + (-1)^{n-1} f(A_{12 \cdots n}),$$

where $A_{rs \cdots z}$ is defined to be $A_r \cap A_s \cap \cdots \cap A_z$, if $r, s, \cdots, z$ are all different, and the summations are over all possible sets of suffixes.

7. If the conditions of Exercise 6 hold, and if also $f(A \cap B) \leq f(A)f(B)$ for all sets A, B, and $f(A) \leq 1$ for all A, prove that

$$f(A_1 \cup A_2 \cup \cdots \cup A_n) \geq \Sigma f(A_r) - \Sigma f(A_r)f(A_s) + \cdots$$
$$+ (-1)^{n-1} f(A_1)f(A_2) \cdots f(A_n).$$

8. Prove the identities

$$A \times (\bigcup_{B \in \mathscr{B}} B) = \bigcup_{B \in \mathscr{B}} (A \times B), \qquad A \times (\bigcap_{B \in \mathscr{B}} B) = \bigcap_{B \in \mathscr{B}} (A \times B).$$

9. Are the sets $A \cup (B \times C)$, $(A \times B) \cup (A \times C)$, $(A \cup B) \times (A \cup C)$ equal, in general? Are any two of the three equal?

10. Show that a set is infinite if and only if it is equivalent to a proper subset of itself. (Prove the result first for countable sets.)

11. Prove that the class of all finite sets of positive integers is countable.

12. Can an uncountable union of (distinct) sets be countable?

13. Use Theorem 1.2d (together with the result which immediately precedes it) to give an alternative proof that the real numbers are uncountable.

14. Exhibit an explicit 1-1 correspondence between (a) $]0, 1[$ and $]0, 1]$; (b) $]0, 1[$ and $[0, 1]$.

15. Show that each of the following sets is a continuum: (a) the unit cube in R^n, (b) the set of all real sequences $x = (x_1, x_2, \cdots)$ where $0 \leq x_r \leq 1$ for all r; (c) the class of all finite sets of real numbers; (d) the real continuous functions on $[0, 1]$; (e) the real step functions on $[0, 1]$ (a step function is a finite linear combination of characteristic functions of intervals).

16. The set of all bounded real functions on $[0, 1]$ is not a continuum.

17. Which points of Cantor's set are not of the form $f(x)$ with $x \in [0, 1]$ (with the notation of Sec. 1.2)? Are the functions f, g, h continuous?

18. Prove int $A \cap$ int $B =$ int $(A \cap B)$, int $A \cup$ int $B \subset$ int $(A \cup B)$. Produce an example with int $A \cup$ int $B \neq$ int $(A \cup B)$. Dualize to closures.

19. In one dimension, show that a bounded open set is uniquely expressible as a countable disjoint union of open intervals.

20. In n dimensions, prove that a bounded open set is expressible as a countable disjoint union of half-open intervals $\{x: a_r \leq x_r < b_r, 1 \leq r \leq n\}$. Show that the result is not true of "half-open" is replaced by "open." Show also that the expression as a union of half-open intervals is never unique (if the set is not empty).

21. With each rational number p/q in $[0, 1]$ associate the open interval $]p/q - \epsilon/q^2,$ $p/q + \epsilon/q^2[$ ($\epsilon > 0$). Show that if ϵ is small enough, these intervals do not form a covering of $[0, 1]$. (Hint: Consider $2^{-1/2}$.)

22. Every covering of a one-dimensional interval by nondegenerate intervals has a countable refinement. Generalize.

23. Is the Heine-Borel theorem true if "bounded closed interval" is replaced by "open interval"? If "open interval" is replaced by "nondegenerate closed interval"?

24. Exhibit a function which is continuous at each point of $]0, 1[$, but not uniformly continuous in $]0, 1[$. If a function has a bounded derivative in an interval I (not necessarily closed), then it is uniformly continuous in I.

25. If f and g are continuous, prove that $f \vee g$ and $f \wedge g$ are also continuous. Generalize to finite families, and show that the results cannot be generalized to countable infinite families.

26. What are the characteristic functions of $A \cup B$, $A \cap B$, $A \setminus B$ in terms of χ_A and χ_B? If (A_i) is a family of sets, what are the characteristic functions of $\bigcup A_i$, $\bigcap A_i$, in terms of the χ_{A_i}?

27. The set A is compact, the real functions f_r ($r = 1, 2, \cdots$) and f are continuous, and $f_r \to f$ pointwise on A. Prove that, if $f_r \to f$ monotonically, then the convergence is uniform on A (Dini's theorem). Show also that the convergence is not uniform in general.

[2]

Lebesgue Measure

2.1. Preliminaries

Measure theory is an attempt to give precise mathematical form to our intuitive idea of what the "size" of a set should be. We require, for example that the measure of a set should not be less than the measure of a subset, and that the measure of a disjoint union should be the sum of the measures of the relevant sets—at least, in the finite case.

Let I be a bounded interval (a, b) in R^n:

$$I = \{x: a_r \prec x_r \prec b_r, 1 \leqslant r \leqslant n\}.$$

The *measure* $m(I)$ of I ($=$ length, area, volume, $\cdots$) is the nonnegative real number defined by

$$m(I) = (b_1 - a_1)(b_2 - a_2) \cdots (b_n - a_n).$$

Thus $m(I) = 0$ if and only if I is degenerate. If I is a nondegenerate unbounded interval, we define $m(I)$ to be $+ \infty$. We do not at present define the measure of a degenerate unbounded interval; it will appear later that this must be taken as zero. (It is a useful, although not entirely reliable, principle in measure theory that $0 \cdot \infty = 0$.)

By the above definition, the measures of the closed interval $[a, b]$, the open interval $]a, b[$, and all intermediate cases are the same. The n-dimensional measure of the boundary of an interval in R^n is thus assumed to be zero, from the beginning. This is, of course, in accordance with our intuitive ideas. There is, however, another more sophisticated approach: If initially we define measure only for half-open intervals, such as $]a, b]$, then it follows as a theorem that the measures of all intervals (a, b) are equal. We do not discuss this here.

The basic problem in measure theory is to extend this notion of size or measure from intervals to some suitably general class of sets. The properties which one wishes this class to have are described later in Sec. 2.5. Some extensions are easy; for instance, to the class $\mathscr{I}$ of sets which are finite unions of intervals. Any such set is evidently a finite disjoint union; and if $A \in \mathscr{I}$ is written as a disjoint union in two ways,

$$A = \bigcup_{r=1}^{p} I_r = \bigcup_{s=1}^{q} I'_s,$$

17

then it is elementary—although rather tedious—to prove that

$$\sum_{r=1}^{p} m(I_r) = \sum_{s=1}^{q} m(I'_s),$$

and the measure of A can then be defined by

$$m(A) = \sum_{r=1}^{p} m(I_r).$$

We shall take for granted any "obvious" properties of the class $\mathscr{I}$ which we require in what follows. In particular, we use (as well as the result just mentioned) the fact that if A, $B \in \mathscr{I}$ and $A \subset B$ then $m(A) \leq m(B)$.

Another extension is obtained by considering the dissection of intervals into congruent subsets; the measure of each such subset is then a known fraction of the measure of the interval. In this way the area of a right-angled triangle in the plane may be found. Classical arguments now lead to the area of any triangle, hence of any polygon.

The next stage in the extension process is brought about by using a device already known to the Greeks (Archimedes, in particular). Suppose A is a plane set; let P and Q be polygons with $P \subset A \subset Q$. Then evidently we should have, for the measure of A,

$$m(P) \leq m(A) \leq m(Q).$$

If, now, there are sequences of polygons P_r, Q_r, with $P_r \subset A \subset Q_r$ for all r, and $m(Q_r) - m(P_r) \to 0$ as $r \to \infty$, it is natural to define $m(A)$ to be the common limit of the sequences $m(P_r)$, $m(Q_r)$. This principle was used to good effect in classical times to calculate the areas of various simple plane sets with curved boundaries. There are, of course, two problems here: the first (logically; second historically) is to decide whether a given set has an area at all, by the above definition, and the second is to calculate it, if it has one.

The above notion of "measure" (usually called "content" here) is essentially that of the Riemann theory of integration. (See Rogosinski's *Volume and Integral*, Chap. II, for a detailed discussion.) The notion has certain deficiencies, in that countable unions and intersections do not always have the properties we would wish. For this reason we have to look for generalizations; the one to be described is essentially due to Lebesgue.

The starting point is the class $\mathscr{J}$ of sets which are countable unions of intervals. This class itself is not adequate for our purposes. For instance, the complement (with respect to an interval) of a set in $\mathscr{J}$ is not necessarily in $\mathscr{J}$. Cantor's set T provides an example of this; it is the complement with respect to [0, 1] of a countable union of intervals, and is not such a union itself. The class $\mathscr{J}$, however, is a very convenient tool in defining the Lebesgue-measurable sets that we want, and we begin by discussing it in some detail.

2.2. The Class $\mathscr{J}$

It is convenient to restrict attention for the present to subsets of some fixed bounded interval X in R^n, which will be the universal set in what follows. It is also convenient to take X to be closed in R^n, and we assume this. The class $\mathscr{J}$ $(=\mathscr{J}(X))$ is then the class of all subsets of X which are countable unions of intervals. If a set is denoted by J, J', J_1, $\cdots$, it is implied that it is in $\mathscr{J}$.

If a set is a finite union of intervals, it is obvious that it can be expressed as a finite disjoint union of intervals. A similar result holds for sets in $\mathscr{J}$:

LEMMA 2.2a. *Any set in $\mathscr{J}$ is a countable union of disjoint intervals.*

Proof. Suppose that $J = \bigcup_{r=1}^{\infty} I_r$. Write $J_1 = I_1$, $J_2 = I_2 \setminus J_1$, $\cdots$; in general, if $J_1, \cdots, J_n$ have been defined, let

$$J_{n+1} = I_{n+1} \setminus (J_1 \cup J_2 \cup \cdots \cup J_n).$$

Then $J = \bigcup_{r=1}^{\infty} J_r$; the sets J_n are disjoint, by construction, and each is a finite union (hence a finite disjoint union) of intervals. It follows that J can be put in the required form.

The expression of a set $J \in \mathscr{J}$ as a disjoint union of intervals is clearly never unique (except in trivial cases); any nondegenerate interval can be written as a union of intervals in infinitely many ways. Fortunately for measure-theoretic purposes, they are all equivalent.

LEMMA 2.2b. *If $J \in \mathscr{J}$ is written as a countable disjoint union of intervals in two ways,*

$$J = \bigcup_{r=1}^{\infty} I_r = \bigcup_{s=1}^{\infty} I'_s,$$

then

$$\sum_{r=1}^{\infty} m(I_r) = \sum_{s=1}^{\infty} m(I'_s).$$

Proof. Both series converge; since all intervals are subsets of X, it follows that

$$\sum_{r=1}^{N} m(I_r) \leq m(X),$$

and similarly for the other series. Suppose that the two sums are different; then there is an integer N such that (say)

$$0 < h = \sum_{r=1}^{N} m(I_r) - \sum_{s=1}^{\infty} m(I'_s).$$

Now choose open intervals $L_s \supset I'_s$ so that $m(L_s) < m(I'_s) + 2^{-s-2}h$; and closed intervals $M_r \subset I_r$ so that $m(M_r) > m(I_r) - 2^{-r-2}h$. It follows that

$$\sum_{r=1}^{N} m(M_r) > \sum_{s=1}^{\infty} m(L_s).$$

This now leads to a contradiction. The set $K = \bigcup_{r=1}^{N} M_r$, being a closed subset of the compact set X, is itself compact (Theorem 1.4e). It is a subset of J, and $\{L_s\}$ forms an open covering of J, hence of K. There is thus a finite refinement, say $\{L_1, L_2, \cdots, L_P\}$. Then

$$\sum_{r=1}^{N} m(M_r) \leq \sum_{s=1}^{P} m(L_s) \leq \sum_{s=1}^{\infty} m(L_s),$$

which contradicts our previous inequality.

COROLLARY. If J is an interval, and $J = \bigcup_{r=1}^{\infty} I_r$ is a representation as a disjoint countable union, then $m(J) = \sum_{r=1}^{\infty} m(I_r)$.

In view of Lemma 2.2b and its corollary, it is now possible to define measure for sets in $\mathscr{J}$; if $J \in \mathscr{J}$ and $J = \bigcup_{r=1}^{\infty} I_r$ is any representation as a countable disjoint union of intervals, the *measure* $m(J)$ of J is

$$m(J) = \sum_{r=1}^{\infty} m(I_r).$$

This is clearly a generalization of measure as previously defined for intervals. It is evident that the empty set $\varnothing$ is in $\mathscr{J}$ and $m(\varnothing) = 0$; more generally, any countable set is in $\mathscr{J}$ and has measure zero.

An immediate consequence of the definition is that if $J \in \mathscr{J}$, $\epsilon > 0$ are given, there is a finite union of intervals $J_0 \subset J$ such that $m(J_0) > m(J) - \epsilon$. This results from the convergence of the series $\sum m(I_r)$.

It should be remarked that in the proof of Lemma 2.2b the topological property of compactness plays a very essential part. In fact, the theory can be developed without reference to compactness (although the approach which we have followed appears the most natural one), for the theory of measure in R can be developed without topological assistance (cf. Burkill, *The Lebesgue Integral*, Chap. II) and transferred to R^n by the device indicated at the end of the chapter (Exercises 16 and 17).

Measure, for sets of $\mathscr{J}$, behaves as we would hope; the next result points in this direction:

LEMMA 2.2c. (1) If $J_1 \subset J_2$, then $m(J_1) \leq m(J_2)$. (2) If $J = \bigcup_{r=1}^{\infty} I_r$, then $m(J) \leq \sum_{r=1}^{\infty} m(I_r)$.

Proof. (1) This is really contained in the proof of Lemma 2.2b above. If $J_1 = \bigcup_{r=1}^{\infty} I_r$, $J_2 = \bigcup_{s=1}^{\infty} I'_s$, where the unions are both disjoint, then the assumption that $m(J_1) > m(J_2)$ leads to a contradiction as before. The result may also be proved otherwise.

(2) Write $I'_1 = I_1, \cdots$; if $I'_1, \cdots, I'_r$ have been defined, write $I'_{r+1} = I_{r+1} \setminus (I'_1 \cup I'_2 \cup \cdots \cup I'_r)$. Then $m(I'_r) \leq m(I_r)$ for all r; $m(J) = \sum_{r=1}^{\infty} m(I'_r) \leq \sum_{r=1}^{\infty} m(I_r)$. It is not, of course, asserted that the series on the right is convergent.

The union of any countable class of sets in $\mathscr{J}$ is evidently again in $\mathscr{J}$. The same is true of any finite intersection. The difference $J_1 \setminus J_2$ of two sets in $\mathscr{J}$ is *not* in general in $\mathscr{J}$. For instance, Cantor's set T is the difference of the closed unit interval and a countable union of open intervals. It is no in $\mathscr{J}$, since it contains no nondegenerate intervals and is not a countable union of points. Likewise, a countable intersection of sets in $\mathscr{J}$ is not in general in $\mathscr{J}$; Cantor's set is the intersection of the countable collection $\{I \setminus E_k\}$, where I is the closed unit interval and E_k is (as defined on p. 6) a finite union of intervals. Each difference is then a finite union of intervals, hence, in $\mathscr{J}$.

LEMMA 2.2d. *If $J_1 \subset J_2 \subset J_3 \subset \cdots$, and $J = \bigcup_{r=1}^{\infty} J_r$, then $m(J) = \lim_{r \to \infty} m(J_r)$.*

Proof. It is clear from the first part of Lemma 2.2c that $\lim m(J_r)$ exists and does not exceed $m(J)$. Let $J_r = \bigcup_{s=1}^{\infty} I_{rs}$; then $J = \bigcup_{r=1}^{\infty} \bigcup_{s=1}^{\infty} I_{rs}$, and this may be arranged as a single sequence, $J = \bigcup_{t=1}^{\infty} I_t$. Write $I'_1 = I_1, \cdots, I'_t = I_t \setminus (I'_1 \cup \cdots \cup I'_{t-1})$. Then, given $\epsilon > 0$, there is an integer N such that

$$m\left(\bigcup_{t=1}^{N} I'_t\right) > m(J) - \epsilon.$$

Also, there is an integer P such that

$$J_P \supset \bigcup_{t=1}^{N} I_t \supset \bigcup_{t=1}^{N} I'_t;$$

hence $m(J_P) > m(J) - \epsilon$. Since ϵ was arbitrary, the result follows.

THEOREM 2.2e. *For any two sets $J, J' \in \mathscr{J}$,*

$$m(J \cup J') + m(J \cap J') = m(J) + m(J').$$

Proof. Suppose that J, J' are finite unions of intervals; then $J' \setminus J$ is also a finite union, and $J \cup J' = J \cup (J' \setminus J)$; so

$$m(J \cup J') = m(J) + m(J' \setminus J).$$

Also, $J' = (J \cap J') \cup (J' \setminus J)$, whence

$$m(J') = m(J \cap J') + m(J' \setminus J);$$

the required result now follows by subtraction.

In the general case $J' \setminus J$ is not necessarily in $\mathscr{J}$, and, in any case, we have not yet established the result (which we assume for *finite* unions of intervals) that the measure of a disjoint union is the sum of the measures of the sets in question. We therefore proceed as follows:

Let $J = \bigcup_{r=1}^{\infty} I_r$, $J' = \bigcup_{s=1}^{\infty} I_s'$; write $J_t = \bigcup_{r=1}^{t} I_r$, $J_t' = \bigcup_{s=1}^{t} I_s'$. Then J_t, J_t', $J_t \cup J_t'$, $J_t \cap J_t'$ are increasing sequences of sets whose unions are, respectively, J, J', $J \cup J'$, $J \cap J'$; and, for each integer t,

$$m(J_t \cup J_t') + m(J_t \cap J_t') = m(J_t) + m(J_t'),$$

as already proved. The general result now follows on making $t \to \infty$, by Lemma 2.2d.

THEOREM 2.2f. If J_1, J_2, $\cdots$ is a countable collection of sets in $\mathscr{J}$, then

$$m(\bigcup J_r) \le \sum m(J_r);$$

and, if the sets J_r are disjoint, then

$$m(\bigcup J_r) = \sum m(J_r).$$

Proof. If there are two sets J_1, J_2 the result

$$m(J_1 \cup J_2) \le m(J_1) + m(J_2)$$

follows at once from Theorem 2.2e, with equality in the case $J_1 \cap J_2 = \varnothing$. The result

$$m(J_1 \cup \cdots \cup J_k) \le m(J_1) + \cdots + m(J_k),$$

with equality in the disjoint case, follows immediately by induction.

In general, let there be infinitely many sets J_1, $J_2 \cdots$; write $J_s' = J_1 \cup J_2 \cup \cdots \cup J_s$. Then the sets J_s' form an increasing sequence whose union is J; by Lemma 2.2d,

$$m(J) = \lim_{s \to \infty} m(J_s') \le \lim_{s \to \infty} \sum_{r=1}^{s} m(J_r) = \sum_{r=1}^{\infty} m(J_r),$$

again with equality in the disjoint case.

In the nondisjoint case the series $\sum m(J_r)$ may, of course, diverge.

The last two theorems of this section are not strictly necessary for the development of the theory. They, however, throw some light on the part played by open sets in some accounts of the theory, and are worth including from this point of view. Theorem 2.2g will also be used in Chap. 6 (in the proof of Theorem 6.1a).

THEOREM 2.2g. *Every open set is in $\mathscr{J}$.*

Proof. Consider all intervals of the form

$$\{x: p_r 2^{-q} \leq x_r \leq (p_r + 1) 2^{-q}, 1 \leq r \leq n\},$$

where $p_1, \cdots, p_n, q$ are integers, with $q > 0$. Let A be the open set, and $\mathscr{J}_q$ the class of all intervals of the above form which are contained in A; this class is evidently finite, since A is a subset of the bounded universal set X. Let A_q be the union of all the intervals in $\mathscr{J}_q$; it is clear, on the one hand, that $A \supset \bigcup_{q=1}^{\infty} A_q$.

On the other hand, if $x \in A$, there is an open interval containing x and contained in A. Hence, if q is large enough, x belongs to some interval of $\mathscr{J}_q$. It follows that $A \subset \bigcup_{q=1}^{\infty} A_q$, hence that $A = \bigcup_{q=1}^{\infty} A_q$, so that $A \in \mathscr{J}$.

In general, a closed set is not in $\mathscr{J}$; for example, Cantor's set T.

THEOREM 2.2h. *If $J \in \mathscr{J}$, there is for each $\epsilon > 0$ an open set $J'(\epsilon) \supset J$, with $m(J'(\epsilon)) < m(J) + \epsilon$.*

Proof. Let $J = \bigcup_{r=1}^{\infty} I_r$, where the intervals I_r are disjoint. For each r there is an open interval $I_r' \supset I_r$, with $m(I_r') < m(I_r) + \epsilon 2^{-r-1}$. Then, if $J' = \bigcup_{r=1}^{\infty} I_r'$, J' is open and

$$m(J') \leq \sum_{r=1}^{\infty} m(I_r') < \sum_{r=1}^{\infty} m(I_r) + \epsilon = m(J) + \epsilon.$$

It is evident that if $J \in \mathscr{J}$ there is an open set $J'' \subset J$ with $m(J'') = m(J)$; let J be represented as a disjoint union of intervals, and let J'' be the union of the interiors of those intervals.

2.3. Measurable Sets

To begin with, we maintain the assumption that all sets are contained in some bounded interval X; this restriction will be removed presently.

The *outer measure* $m^*(A)$ of the set A is defined by

$$m^*(A) = \inf_{A \subset J} m(J).$$

The *inner measure* $m_*(A)$ is defined by

$$m_*(A) = m(X) - m^*(X \setminus A).$$

The definition of inner measure can be given in a slightly different form. The measure of a set K whose complement (with respect to X) is in $\mathscr{J}$ can be defined by $m(K) = m(X) - m(X \setminus K)$; such sets have, of course, properties dual to those of the class $\mathscr{J}$.

It is then clear that

$$m_*(Z) = \sup_{K \subset A} m(K)$$

where the supremum is over all subsets K of A whose complements are in $\mathscr{J}$.

It might have been expected that the inner measure of A would have been defined to be $\sup m(J)$ for $J \subset A$. It may be worth while to explain just why this is not appropriate. What is wanted, of course, is a good outer approximation to the set A by sets whose measures are known; and also a good inner approximation. As far as outer approximation is concerned, there is no advantage in considering sets K whose complements are in $\mathscr{J}$, as well as sets in $\mathscr{J}$. Suppose $A \subset K$, where $X \setminus K \in \mathscr{J}$; for any given $\epsilon > 0$ there is a finite union of intervals, say J', contained in $X \setminus K$, such that $m(J') > m(X \setminus K) - \epsilon$. The set $J = X \setminus J'$ is also a finite union of intervals, and contains A; and $m(J) < m(K) + \epsilon$. It follows that $\inf_{A \subset J} m(J) \leq \inf_{A \subset K} m(K)$; that is, sets $J \in \mathscr{J}$ always give at least as good an approximation as sets K with $X \setminus K \in \mathscr{J}$. It may happen that sets K fail to give as good an approximation as sets J. Take the subset A of $[0, 1]$ consisting of all rational numbers; $A \in \mathscr{J}$ and $m(A) = 0$. If $A \subset K$, then the complement of K consists of irrational numbers only, hence contains no nondegenerate interval, hence is of measure zero. The measure of K is thus 1. The dual results hold for inner approximation; sets K are at least as effective as sets J and may, in some cases, be more effective. It may happen that any set J contained in A has a measure which is much less than the measure of A "ought" to be; consider the set A of irrational points of $[0, 1]$.

On the other hand, in order to obtain a good outer approximation to a set, it is not necessary to consider the whole class $\mathscr{J}$. In view of Theorems 2.2g and 2.2h, it would be sufficient to take the open sets; $m^*(A) = \inf m(G)$ for all open sets $G \supset A$. Dually, $m_*(A) = \sup m(H)$ for compact subsets H of A. These facts may be used to give alternative proofs of some of the theorems which follow.

A bounded set may be regarded as a subset of infinitely many bounded intervals X; but the choice of X does not affect inner and outer measure:

LEMMA 2.3a. *The inner and outer measures of a set are independent of the interval X.*

Proof. This is almost obvious in the case of outer measure. To verify the result for inner measure, suppose X and X' are two bounded intervals containing A; without loss of generality, it may be supposed that $X' \supset X$. If $X \supset J \supset X \setminus A$, then $J' = J \cup (X' \setminus X)$ is in $\mathscr{J}$, and $X' \supset J' \supset X' \setminus A$. Also, if $X' \supset J' \supset X' \setminus A$, the set $J = J' \cap X$ is in $\mathscr{J}$, and $X \supset J \supset X \setminus A$. In either case $J' = J \cup (X' \setminus X)$, whence

$$m(J') = m(J) + m(X' \setminus X) = m(J) + m(X') - m(X),$$

which implies that

$$\inf_{J' \supset X' \setminus A} m(J') = \inf_{J \supset X \setminus A} m(J) + m(X') - m(X);$$

that is,

$$m(X') - m^*(X' \setminus A) = m(X) - m^*(X \setminus A),$$

which is the result required.

It is obvious that if $A' \supset A$ then $m^*(A') \geq m^*(A)$ and $m_*(A') \geq m_*(A)$. It is almost obvious that inner and outer measure are related in the way one would expect:

LEMMA 2.3b. For any set A, $m^*(A) \geq m_*(A)$.

Proof. Let J, J' be subsets of X containing A, $X \setminus A$ respectively; then $J \cup J' = X$, and (by Theorem 2.2f) $m(X) \leq m(J) + m(J')$; that is, $m(J) \geq m(X) - m(J')$. It follows that $m^*(A) = \inf m(J) \geq \sup (m(X) - m(J')) = m(X) - \inf m(J') = m(X) - m^*(X \setminus A) = m_*(A)$.

LEMMA 2.3c. If $J \in \mathscr{J}$, then $m^*(J) = m(J) = m_*(J)$.

Proof. It is trivial that $m^*(J) = m(J)$. Also, given $\epsilon > 0$, there is a finite union of intervals $J' \subset J$ with $m(J') > m(J) - \epsilon$. Then $X \setminus J'$ is a finite union of intervals and contains $X \setminus J$, and $m(X) - m(X \setminus J') > m(J) - \epsilon$. Hence $m_*(J) \geq m(J) - \epsilon$; since ϵ is arbitrary, $m_*(J) \geq m(J)$. Taking into account Lemma 2.3b, the required result now follows.

We now define a set E to be *measurable* (more accurately, measurable in the sense of Lebesgue or Lebesgue measurable) if its inner and outer measures are equal; the *measure* $m(E)$ of E is then defined to be the common value of $m^*(E)$ and $m_*(E)$. It follows from Lemma 2.3c that this generalizes measure as defined for sets in $\mathscr{J}$; it will appear later (Sec. 2.5) that the generalization is a real one; there are sets which are measurable but which are not in $\mathscr{J}$ and whose complements are not in $\mathscr{J}$.

One immediate consequence of the definition just given is that the set E is measurable if and only if its complement $X \setminus E$ is measurable; and, if so, $m(E) + m(X \setminus E) = m(X)$.

Before proceeding to develop the main properties of measurable sets, it is convenient to remove the restriction that they should be bounded. Let A be a set which is not necessarily bounded; its *outer measure* $m^*(A)$ is defined to be sup $m^*(B)$ for all bounded subsets B of A, and its *inner measure* $m_*(A)$ is sup $m_*(B)$ for the same class of sets B. It is obvious that these definitions reduce to those previously given if A is bounded.

The definitions can be put in several equivalent forms. For example, it is clear that it would be enough to consider only those bounded subsets B which are of the form $A \cap I$, where I is a bounded interval. Specializing even more, let $I^{(k)} = \{x: -k \leq x_r \leq k, 1 \leq r \leq n\}$ (where k is a positive integer), and let $A^{(k)} = A \cap I^{(k)}$. Then, clearly, $m^*(A) = \sup_k m^*(A^{(k)}) = \lim_{k \to \infty} m^*(A^{(k)})$, and $m_*(A) = \sup_k m_*(A^{(k)}) = \lim_{k \to \infty} m_*(A^{(k)})$.

It turns out that for unbounded sets the requirement that the inner and outer measures should be equal is not sufficient to ensure good behavior; there are sets with both inner and outer measure equal to ∞ which have undesirable properties. So we define a set E to be *measurable* if $E^{(k)}$ is measurable for every k; and then the *measure* of E is defined by

$$m(E) = \lim_{k \to \infty} m(E^{(k)}).$$

This clearly reduces to the definition already given, if E is bounded.

The measure of an unbounded measurable set may be finite or infinite; the measure of a bounded measurable set is, of course, finite. It is clear that a set is measurable if and only if its complement (with respect to the whole space) is measurable.

If E is measurable, then $m^*(E)$ and $m_*(E)$ are equal; if $m^*(E)$ and $m_*(E)$ are finite and equal, then it can be shown (see the exercises at the end of this chapter) that E is measurable and $m(E) = m^*(E) = m_*(E)$.

There are various other ways of defining measure for sets which are not necessarily bounded. In particular, one may first define measure for sets which are countable unions of intervals, but are not necessarily bounded, in very much the same way as for sets in $\mathscr{J}$. One may then define outer measure directly, and, by considering complements of such sets, inner measure also. Measurability and measure are then defined in the obvious way, with the appropriate reservations in the case where outer and inner measure are both infinite.

It is obvious that the conclusion of Lemma 2.3b, that $m^*(A) \geq m_*(A)$ for any set A, holds also for unbounded sets; and that if $A' \supset A$, then $m^*(A') \geq m^*(A)$ and $m_*(A') \geq m_*(A)$. It is also clear that if E and E' are measurable, and $E' \supset E$, then $m(E') \geq m(E)$.

LEMMA 2.3d. If A and A' are any sets, then

(1) $m^*(A \cup A') + m^*(A \cap A') \leq m^*(A) + m^*(A')$;

(2) $m_*(A \cup A') + m_*(A \cap A') \geq m_*(A) + m_*(A')$.

Proof. (1) Suppose that A and A' are bounded; take any suitable interval X containing them both. Given $\epsilon > 0$, there are sets $J \supset A$, $J' \supset A'$, with $m(J) < m^*(A) + \epsilon$, $m(J') < m^*(A') + \epsilon$.

Then $J \cup J' \supset A \cup A'$, $J \cap J' \supset A \cap A'$, and

$$m^*(A \cup A') + m^*(A \cap A') \leq m(J \cup J') + m(J \cap J')$$
$$= m(J) + m(J')$$

(by Theorem 2.2e)

$$< m^*(A) + m^*(A') + 2\epsilon,$$

from which the result follows, since ϵ is arbitrary. To extend the result to unbounded sets, consider intersections with the interval $I^{(k)}$; we have $(A \cup A')^{(k)} = A^{(k)} \cup A'^{(k)}$, $(A \cap A')^{(k)} = A^{(k)} \cap A'^{(k)}$, and

$$m^*(A^{(k)} \cup A'^{(k)}) + m^*(A^{(k)} \cap A'^{(k)}) \leq m^*(A^{(k)}) + m^*(A'^{(k)}),$$

by the result just proved; the required relation now follows on making $k \to \infty$.

(2) Again suppose first that A and A' are subsets of X; then

$$m_*(A \cup A') + m_*(A \cap A') = 2m(X) - m^*(X \setminus (A \cup A')) - m^*(X \setminus (A \cap A'))$$
$$= 2m(X) - m^*(X \setminus A \cap X \setminus A')$$
$$- m^*(X \setminus A \cup X \setminus A')$$
$$\geq 2m(X) - m^*(X \setminus A) - m^*(X \setminus A')$$

(by the first part of the lemma)

$$= m_*(A) + m_*(A').$$

The result for unbounded sets follows as in (1).

LEMMA 2.3e. If $A = \bigcup_{r=1}^{\infty} A_r$, then $m^*(A) \leq \Sigma_{r=1}^{\infty} m^*(A_r)$.

Proof. If all sets are contained in a bounded interval X, then, for any given $\epsilon > 0$, and each integer r, there is a set $J_r \supset A_r$ with $m(J_r) < m^*(A_r) + \epsilon 2^{-r-1}$. Then if $J = \bigcup J_r$, $J \supset A$ and $m^*(A) \leq m(J)$; by Theorem 2.2f, $m(J) \leq \Sigma m(J_r) \leq \Sigma m^*(A_r) + \epsilon$. The result now follows in the bounded case. To establish the general case, note that $m^*(A^{(k)}) \leq \Sigma m^*(A_r^{(k)})$, by the result just proved; this is obviously not greater than $\Sigma m^*(A_r)$, and the theorem follows on making $k \to \infty$.

It is now possible to prove that measurable sets have all the properties which are required for integration theory (and which are not all enjoyed by sets in $\mathscr{J}$, or their complements).

THEOREM 2.3f. If E and E' are measurable, so are $E \cup E'$ and $E \cap E'$; and

$$m(E \cup E') + m(E \cap E') = m(E) + m(E').$$

Proof. Suppose that the sets are bounded. It follows from Lemma 2.3d that

$$
\begin{aligned}
m(E) + m(E') &\geq m^*(E \cup E') + m^*(E \cap E') \\
&\geq m_*(E \cup E') + m_*(E \cap E') \\
&\geq m(E) + m(E').
\end{aligned}
$$

Hence $m^*(E \cup E') + m^*(E \cap E') = m_*(E \cup E') + m_*(E \cap E')$; and, since the inner measure of a set cannot exceed its outer measure, it follows at once that $m^*(E \cup E') = m_*(E \cup E')$ and $m^*(E \cap E') = m_*(E \cap E')$, so that $E \cup E'$ and $E \cap E'$ are both measurable. The relation $m(E \cup E') + m(E \cap E') = m(E) + m(E')$ is an immediate consequence of Lemma 2.3d.

If the sets are unbounded, the first part of the proof shows that $(E \cup E')^{(k)}$ and $(E \cap E')^{(k)}$ are measurable for all sufficiently large k, so that $E \cup E'$ and $E \cap E'$ are measurable. Also,

$$
m((E \cup E')^{(k)}) + m((E \cap E')^{(k)}) = m(E^{(k)}) + m(E'^{(k)}),
$$

and the required relation follows from this on making $k \to \infty$.

COROLLARY. *If E and E' are measurable, so is $E \setminus E'$.*

Proof. $E \setminus E' = E \cap \mathscr{C}E'$, and $\mathscr{C}E'$ is measurable.

THEOREM 2.3g. *Any countable union or countable intersection of measurable sets is measurable. In general, $m(\bigcup E_r) \leq \sum m(E_r)$, with equality if the sets E_r are disjoint.*

Proof. If there are two sets, the assertions of the theorem are immediate consequences of Theorem 2.3f, and the case of a finite number of sets follows at once by induction.

To extend the results to infinite unions and intersections, suppose first that the sets E_r are disjoint and are all contained in a bounded interval X. By Lemma 2.3e,

$$
m^*\left(\bigcup_{r=1}^{\infty} E_r\right) \leq \sum_{r=1}^{\infty} m^*(E_r) = \sum_{r=1}^{\infty} m(E_r).
$$

Also, for any positive integer p,

$$
m_*\left(\bigcup_{r=1}^{\infty} E_r\right) \geq m_*\left(\bigcup_{r=1}^{p} E_r\right) = m\left(\bigcup_{r=1}^{p} E_r\right) = \sum_{r=1}^{p} m(E_r),
$$

so that

$$
m_*\left(\bigcup_{r=1}^{\infty} E_r\right) \geq \sum_{r=1}^{\infty} m(E_r) \geq m^*\left(\bigcup_{r=1}^{\infty} E_r\right),
$$

which shows that $\bigcup_{r=1}^{\infty} E_r$ is measurable, and $m(\bigcup_{r=1}^{\infty} E_r) = \sum_{r=1}^{\infty} m(E_r)$.

If the sets E_r are disjoint, but not necessarily uniformly bounded, observe that $E_r^{(k)} = E_r \cap I^{(k)}$ is measurable for all r and all k, and hence $(\bigcup E_r)^{(k)} = \bigcup E_r^{(k)}$ is measurable for all k, which implies that $\bigcup E_r$ is measurable. Then $m(\bigcup E_r) \geq m(\bigcup_{r=1}^{p} E_r) = \Sigma_{r=1}^{p} m(E_r)$, for any p, so that $m(\bigcup E_r) \geq \Sigma_{r=1}^{\infty} m(E_r)$. On the other hand,

$$m((\bigcup E_r)^{(k)}) = \sum_{r=1}^{\infty} m(E_r^{'(k)}) \leq \sum_{r=1}^{\infty} m(E_r),$$

so that $m(\bigcup E_r) \leq \Sigma_{r=1}^{\infty} m(E_r)$, giving $m(\bigcup E_r) = \Sigma m(E_r)$ in this case also.

If the sets E_r are not necessarily disjoint, write $E_1' = E_1, \cdots, E_r' = E_r \setminus (E_1 \cup \cdots \cup E_{r-1}), \cdots$. Then each set E_r' is measurable, being the difference of two measurable sets. Hence $\bigcup E_r = \bigcup E_r'$ is measurable, and $m(\bigcup E_r) = \Sigma m(E_r') \leq \Sigma m(E_r)$.

Finally, since $\bigcap E_r = \mathscr{C}(\bigcup \mathscr{C} E_r)$, and a set is measurable if and only if its complement is measurable, any countable intersection of measurable sets is measurable.

It is convenient to include the next theorem at this point; it is an easy consequence of Theorem 2.3g:

THEOREM 2.3h. If the sets E_r are all measurable, if $E_1 \subset E_2 \subset E_3 \subset \cdots$ and $E = \bigcup E_r$, then

$$m(E) = \lim m(E_r);$$

more generally, if $A_1 \subset A_2 \subset A_3 \subset \cdots$ and $A = \bigcup A_r$, then

$$m^*(A) = \lim m^*(A_r).$$

Proof. The measurable case follows at once from Theorem 2.3g. Since $E = E_1 \cup (E_2 \setminus E_1) \cup (E_3 \setminus E_2) \cup \cdots$, it is apparent that $m(E) = \lim (m(E_1) + m(E_2 \setminus E_1) + \cdots + m(E_r \setminus E_{r-1}))$; since evidently $m(E_2 \setminus E_1) = m(E_2) - m(E_1)$, etc., $m(E) = \lim m(E_r)$.

In general, it is clear that, since $A \supset A_r$ for all r, $m^*(A) \geq m^*(A_r)$ for all r, and so $m^*(A) \geq \lim m^*(A_r)$. To establish the opposite inequality, suppose first that the sets A_r are uniformly bounded (all in some bounded interval X). Given $\epsilon > 0$, choose for each r a set $J_r \supset A_r$ with $m(J_r) < m^*(A_r) + \epsilon$. Let $H_r = J_r \cap J_{r+1} \cap J_{r+2} \cap \cdots$; then H_r is measurable (by Theorem 2.3g), $H_r \supset A_r$, and $m(H_r) < m^*(A_r) + \epsilon$. Also, the sets H_r are increasing: $H_1 \subset H_2 \subset H_3 \subset \cdots$. Since $A = \bigcup A_r \subset \bigcup H_r$, we have $m^*(A) \leq m(\bigcup H_r) = \lim m(H_r)$ (by the first part of the theorem), and this does not exceed $\lim m^*(A_r) + \epsilon$. Since ϵ is arbitrary, the equality follows. If the sets A_r are not uniformly bounded, we have $m^*(A^{(k)}) = \lim m^*(A_r^{(k)}) \leq \lim m^*(A_r)$, so the general result is obtained by making $k \to \infty$.

The next theorem is about product sets; E is in R^n, E' in R^p, and $E \times E'$ in R^{n+p}. We use the same notation for measure in the three spaces; this should cause no confusion.

THEOREM 2.3i. If E and E' are measurable, so is $E \times E'$; and $m(E \times E') = m(E)\, m(E')$.

Proof. It is evidently sufficient to prove the theorem for bounded sets; suppose $E \subset X$, $E' \subset X'$. For intervals the theorem is trivial. For sets in $\mathscr{J}$ it is immediate; if $J = \bigcup I_r$, $J' = \bigcup I'_s$, where the unions are both disjoint, then $J \times J' = \bigcup \bigcup I_r \times I'_s$ is a representation of $J \times J'$ as a countable disjoint union of intervals and

$$m(J \times J') = \sum \sum m(I_r \times I'_s) = \sum \sum m(I_r)\, m(I'_s) =$$
$$\sum m(I_r) \sum m(I'_s) = m(J)\, m(J').$$

In general, let $J \supset E$, $J' \supset E'$; then $J \times J' \supset E \times E'$, and so $m^*(E \times E') \le m(J \times J') = m(J)\, m(J')$; since $m(J)$, $m(J')$ can be arbitrarily close to $m(E)$, $m(E')$ respectively, $m^*(E \times E') \le m(E)\, m(E')$. Also, $m_*(E \times E') = m(X \times X') - m^*(X \times X' \setminus E \times E')$, and

$$X \times X' \setminus E \times E' = X \times (X' \setminus E') \cup (X \setminus E) \times E',$$

so that $m^*(X \times X' \setminus E \times E') \le m^*(X \times (X' \setminus E')) + m^*(X \setminus E) \times E'))$, by Lemma 2.3d (1). Hence

$$m^*(X \times X' \setminus E \times E') \le m(X)\, m(X' \setminus E') + m(X \setminus E)\, m(E')$$

and $m(X \setminus E) = m(X) - m(E)$, $m(X' \setminus E') = m(X') - m(E')$, so that

$$m^*(X \times X' \setminus E \times E') \le m(X)\, m(X') - m(E)\, m(E'),$$

and so

$$m_*(E \times E') \ge m(E)\, m(E'),$$

which shows that $E \times E'$ is measurable and has measure $m(E)\, m(E')$.

There are also theorems on the reverse direction, relating the measurability of a section of a set in a product space to the measurability of the set. We prove one such result later (Theorem 4.2a).

The result of the last theorem of this section is to be expected; if it were not so, then Lebesgue measure would fail to satisfy our intuitive ideas of what the measure of a set should be.

THEOREM 2.3j. Measurability and measure are invariant under translations and rotations.

Proof. It is evidently enough to establish the theorem for bounded intervals —indeed, for half-open intervals. For translations the result is immediate. For rotations a direct argument can be given, but it seems better to proceed indirectly, as follows.

In the first place, an open set remains open under rotations, as is obvious from the definition of open sets in terms of spheres. Any interval is the intersection of a countable class of open sets (in fact, of open intervals). Hence a rotated interval (which is not, in general, an interval as defined in Sec. 1.3) is a countable intersection of open sets, and so is measurable, by Theorems 2.2g and 2.3g.

Let I be a bounded half-open interval $\{x: a_r \leq x_r < b_r, 1 \leq r \leq n\}$. Just as in Theorem 2.2g, if one considers intervals of the form

$$\{x: (b_r - a_r) \, p_r 2^{-q} \leq x_r < (b_r - a_r) \, (p_r + 1) \, 2^{-q}\},$$

it is seen that any open set is a countable disjoint union of intervals similar to I. In particular, the interior S of the unit sphere is such a union, say $S = \bigcup I_k$.

Let ρ be a rotation, and ρA the image of the set A under ρ. There is no loss of generality (given invariance under translations) in considering only rotations about the origin. Suppose that I' is an interval similar to I: say $I' = a + pI$. Then $\rho I' = \rho a + p\rho I$, and $m(\rho I') = p^n m(\rho I)$. But $m(I') = p^n m(I)$, and so

$$m(\rho I') = K m(I'),$$

where $K \, (= m(\rho I)/m(I))$ depends on ρ and I but not on I'. By Theorem 2.3g, $m(S) = \sum m(I_k)$, and clearly $m(\rho S) = m(S)$. But

$$m(\rho S) = \sum m(\rho I_k) = K \sum m(I_k) = K m(S).$$

Hence $K = .1$, for any ρ and any I, which proves that the measure of an interval is invariant under rotations.

2.4. Sets of Measure Zero

Since, for any set A, $0 \leq m_*(A) \leq m^*(A)$, a necessary and sufficient condition that E should be (measurable and) of measure zero is that $m^*(E) = 0$. It is clear that any subset of a set of measure zero is itself of measure zero. As a special case of Theorem 2.3g, the union of a countable class of sets of measure zero is of measure zero; a direct proof of this is quite easy. A set of measure zero is sometimes called a null set; we shall not, however, use this term.

A property which holds except on a set of measure zero is said to hold *almost everywhere* or to hold *essentially*.

Thus, if f is the function defined by

$$f(x) = 0 \quad \text{if } x \text{ is irrational}$$
$$f(x) = 1 \quad \text{if } x \text{ is rational,}$$

then $f(x) = 0$ almost everywhere. This is usually written as $f(x) = 0$ p.p. (*presque partout*), or $f(x) =^0 0$. Sometimes, if it is clear from the context that "almost everywhere" is implied, we may omit any explicit indication and write, for example, simply $f(x) = 0$. The terminology may be modified in various trivial ways; for instance, we say that *almost all* points in a set have a certain property if all except those in a set of measure zero have it.

For some purposes (integration theory, in particular) the values taken by a function may be altered on a set of measure zero, without affecting its properties very much. It is convenient to define a function f to be *essentially bounded* on a set E if there is a subset E_0 of E, of measure zero, such that f is bounded on $E \setminus E_0$. The *essential supremum* (essential least upper bound) of f on E is defined by

$$\operatorname*{ess\,sup}_{x \in E} f(x) = \inf \left(\sup_{x \in E \setminus E_0} f(x) \right),$$

the infimum being taken over all subsets E_0 of E which are of measure zero. For some purposes this gives a more accurate bound than the usual supremum; and similarly of course for the essential infimum and infimum. As an example, the function f defined by

$$f(x) = x \quad \text{if} \quad x \text{ is irrational}$$

$$f(x) = n \quad \text{if} \quad x = \frac{m}{n} \text{ is rational}$$

is essentially bounded on $[0, 1]$, and its essential supremum there is 1. In general, two functions f and g are *equivalent* (on E) if $f = g$ p.p. on E; thus f is essentially bounded if it is equivalent to a bounded function, and its essential supremum is the infimum of the suprema of the functions to which it is equivalent.

As has already been noted in Sec. 2.2, the measure of any countable set is necessarily zero. A set of measure zero is not necessarily countable; Cantor's set T, defined in Sec. 1.2, is not countable but is of measure zero. To see this, note that the measure of the set $E_1 \cup E_2 \cdots \cup E_r$ (with the notation used in Sec. 1.2) is $3^{-1}(1 + 2/3 + \cdots + (2/3)^{r-1})$, and so the measure of its complement is $(2/3)^r$. The outer measure $m^*(T)$ of T is less than this, for all r, and is therefore zero.

The rest of this section constitutes a digression, and the results will not be used in the rest of the book; they are, however, of some interest and seem worth giving.

Given any positive integer k, the fractional part of a real number may be written in the form (k-ary representation)

$$x_1 k^{-1} + x_2 k^{-2} + x_3 k^{-3} + \cdots;$$

here each x_r is an integer between 0 and $k - 1$, inclusive. The case $k = 10$ gives the ordinary decimal expansion; the cases $k = 2$ and $k = 3$ have been mentioned in Sec. 1.2. The representation is unique except for certain rational numbers for which two expansions are possible; this exceptional set depends on k, and is always of measure zero (being countable). We may ignore the exceptional cases in what follows. To say that the fractional parts of almost all real numbers have a certain property is evidently equivalent to saying that almost all numbers in (0, 1) have it; we can therefore restrict attention to the unit interval.

The assertion that Cantor's set is of measure zero is equivalent to the assertion that the set of numbers whose ternary expansion (that is, with $k = 3$) does not contain the integer 1 is of measure zero. More generally, it is easy to see that, for any k, the set of real numbers whose expansions fail to contain a given integer between 0 and $k - 1$ is of measure zero, for the set of numbers which do not have the given integer in the first p terms of the expansion is of measure $(k - 1)^p / k^p$, and the outer measure of the required set is less than this for every p. Also, the set of numbers which fail to contain a given finite sequence of integers in their k-ary representation is of measure zero; one has simply to consider k^N instead of k in the above, where N is the number of terms in the sequence. In particular, for any N, almost all numbers have "runs" of N consecutive integers h in the r k-ary representation (where h is any integer between 0 and $k - 1$). As a corollary, almost all numbers have arbitrarily long runs of every possible integer in their representation; the exceptional set is the union of all the sets of numbers which fail to contain a run of more than N integers h; each such set is of measure zero, and the class of such sets is countable.

In a slightly different direction, one would expect that, for most numbers, each integer from 0 to $k - 1$ would occur in the k-ary representation about the same number of times. This is indeed true, in the sense of Theorem 2.4a, which follows.

Let a number x, an integer k, and an integer h between 0 and $k - 1$ be given. Let M ($= M(x, k, h, N)$) be the number of times the integer h occurs in the first N terms of the k-ary representation of x. If, for each possible h, the ratio M/N tends to k^{-1} as $N \to \infty$, the number x is *simply k normal*.

THEOREM 2.4a (E. Borel). Almost all numbers are simply k normal for every k.

Proof. It is enough to prove that for a given fixed k almost all numbers are simply k normal. Now, the set of numbers x for which M/N has a given value

L/N has evidently measure m_L given by

$$m_L = \frac{N!}{L!\,(N-L)!} \cdot \frac{(k-1)^{N-L}}{k^N}.$$

It is an easy elementary computation to show that

$$\sum_{L=0}^{N} (L/N - 1/k)^4\, m_L = N^{-3}\,k^{-2}\,(3(k-1)^2\,N + k^3 - 7k^2 + 12k - 6) < KN^{-2}$$

where K depends on k but not on N. If now $\epsilon > 0$ and

$$E(N, \epsilon) = \{x\colon |M/N - 1/k| > \epsilon\},$$

it is clear that $\sum (L/N - 1/k)^4 m_L > \epsilon^4 m(E(N, \epsilon))$, and hence that

$$m(E(N, \epsilon)) < KN^{-2}\epsilon^{-4}.$$

Let $F(N, \epsilon) = \bigcup_{P=N}^{\infty} E(P, \epsilon)$ and $G(\epsilon) = \bigcap_{N=1}^{\infty} F(N, \epsilon)$.

To say that x is not simply k normal is to say that there is an $\epsilon > 0$ such that $x \in E(N, \epsilon)$ for infinitely many integers N. This is equivalent to the condition that $x \in F(N, \epsilon)$ for all N, and hence to the condition

$$x \in G(\epsilon)$$

for some ϵ. If $\epsilon_1, \epsilon_2, \cdots$ is a sequence of positive numbers tending to zero, x is not simply k normal if and only if

$$x \in \bigcup_{r=1}^{\infty} G(\epsilon_r).$$

The sets F and G are evidently measurable, by Theorem 2.3g; and

$$m(F(N, \epsilon)) \leq \sum_{P=N}^{\infty} m(E(P, \epsilon)) \leq K\epsilon^{-4} \sum_{P=N}^{\infty} P^{-2} \leq K\epsilon^{-4}/(N-1),$$

so that $m(F(N, \epsilon)) \to 0$ as $N \to \infty$; and $m(G(\epsilon)) \leq m(F(N, \epsilon))$ for all N, so that $m(G(\epsilon)) = 0$, for any $\epsilon > 0$. Since each set $G(\epsilon_r)$ is of measure zero, so is their union, which is the set of numbers which are not simply k normal.

A number x is k *normal* if $k^p x$ is simply k^q normal for $p = 0, 1, 2, \cdots$; $q = 1, 2, 3, \cdots$. This is equivalent to requiring that each finite sequence of integers should appear with its expected frequency in the k-ary representation of x. The condition is obviously more restrictive than simple k normality. It follows, however, at once from Theorem 2.4a that almost all numbers are k normal for every k.

2.5. Borel Sets and Nonmeasurable Sets

The class of Lebesgue-measurable sets is larger than is strictly necessary for a Lebesgue-type theory, with satisfactory theorems on limiting processes. We require a class of sets which is closed under countable unions and intersections, and differences. More precisely, we want a class $\mathscr{F}$ of sets such that

(1) if F_1, F_2, $\cdots \in \mathscr{F}$, then $\bigcup_{r=1}^{\infty} F_r \in \mathscr{F}$;

(2) if F_1, $F_2 \in \mathscr{F}$, then $F_1 \setminus F_2 \in \mathscr{F}$.

From (1) and (2) it follows at once that any countable intersection of sets in $\mathscr{F}$ is in $\mathscr{F}$. A class satisfying (1) and (2) is called a σ ring (sigma ring). Naturally, the σ rings of interest in the present context must contain the intervals.

Given any class of sets, there is certainly a σ ring containing it; for instance, the σ ring of all subsets of the appropriate space. The intersection of all σ rings containing the given class is clearly again a σ ring, the smallest which contains the class; it is the σ ring generated by the class in question. The Borel sets (or Borel-measurable sets) in R^n are the sets of the σ ring generated by the intervals in R^n. Since the Lebesgue-measurable sets form a σ ring containing the intervals (Theorem 2.3f, Corollary, and Theorem 2.3g), every Borel set is Lebesgue measurable.

Not every Lebesgue-measurable set is a Borel set. This is most easily established by showing that there are essentially more measurable sets than Borel sets. (An alternative proof is indicated in the exercises at the end of Chap. 3.) It can be proved without difficulty that the Borel sets form a continuum. (The proof involves a slight acquaintance with ordinal numbers and will not be attempted here.) On the other hand, there are as many measurable sets as there are subsets of a continuum; the proof of Theorem 1.2d shows that this is not a continuum. To see that there are at least as many measurable sets as there are subsets of $(0, 1)$, recall that there is a 1-1 map of $(0, 1)$ on to a subset of the Cantor set T (the function f described at the end of Sec. 1.2). Hence there is a 1-1 correspondence between all subsets of $(0, 1)$ and some subsets of T; and each subset of T is measurable, being of measure zero.

Although from the point of view of set-theoretic equivalence there are many more measurable sets than Borel sets, from the point of view of measure theory there are not many more. More precisely, given any measurable set E, there are Borel sets F and G such that $F \subset E \subset G$ and $m(E \setminus F) = m(G \setminus E) = 0$ (which implies $m(F) = m(E) = m(G)$). To prove this, it is evidently enough to consider the bounded case. If J_r is a sequence of sets in $\mathscr{J}$, containing E, such that $m(J_r) \to m(E)$, then $G = \bigcap J_r$ clearly has the required properties; and F is obtained dually. There is, in general, no unique smallest Borel set containing a given set.

Finally, we note that every open set (and hence every closed set) is a Borel set. For bounded open sets this has been established in Theorem 2.2g;

the unbounded case can either be proved directly, on the same lines, or can be deduced from the bounded case by noting that any open set is a countable union of bounded open sets.

Although the class of measurable subsets of R^n is equivalent to the class of all subsets, it can be shown that not every subset of R^n is measurable. This is not a defect in the Lebesgue theory which would disappear in some refinement; it is inherent in the essential structure. Indeed, it will appear below that if we have any definition of measure such that an interval has its usual measure and

(1) if $E_1, E_2, \cdots$ are measurable and disjoint then

$$m \left(\bigcup_{r=1}^{\infty} E_r \right) = \sum_{r=1}^{\infty} m(E_r);$$

(2) if E is measurable then so is $x + E$ for all x, and $m(x + E) = m(E)$ ($x + E = \{z : z = x + y, y \in E\}$)

then not all sets can be measurable. Lebesgue measure evidently enjoys properties (1) and (2); (1) by Theorem 2.3g, and (2) because the measure of an interval, hence of any set in $\mathscr{J}$, has this property (Theorem 2.3j).

We proceed to show the existence of a nonmeasurable subset of the interval [0, 1]. Let ξ be any irrational number; define x and y to be *equivalent* if there are integers p and q (positive, negative, or zero) such that

$$x - y = p\xi + q.$$

This induces a decomposition of [0, 1] into disjoint equivalence classes, all numbers in any one class being equivalent and no two numbers from different classes being equivalent. Let A_0 be any set containing exactly one point from each equivalence class.

For any integer N, let A_N be the set of points of [0, 1] which are of the form $y + N\xi + q$, where $y \in A_0$ and q is an integer. If η is the (unique) point of [0, 1] of the form $N\xi + q$, then it is apparent that the part of A_N between 0 and η is obtained by translating through a distance $1 - \eta$ the part of A_0 between $1 - \eta$ and 1. Similarly, the part of A_N between η and 1 is obtained by translating through a distance η the part of A_0 between 0 and $1 - \eta$. It follows that if A_0 is measurable, so is A_N for all N, and $m(A_N) = m(A_0)$.

The sets A_N are disjoint, for, if $M \neq N$, and $y + M\xi + q = y' + N\xi + q'$, then if $y = y'$ we have $\xi = (q' - q)/(M - N)$, a contradiction, since ξ is irrational. If $y \neq y'$ then $y - y' = (N - M)\xi + q' - q$, so that y and y' are equivalent, contradicting the choice of A_0. And the union of the sets A_N is the interval [0, 1]; for if $x \in [0, 1]$ then x is equivalent to some $x' \in A_0$; $x = x' + N\xi + q$ for some N, and so $x \in A_N$.

If, then, the set A_0 and hence all the sets A_N were measurable, we would have

$$m([0, 1]) = m \left(\bigcup_{N=-\infty}^{+\infty} A_N \right) = \sum_{N=-\infty}^{+\infty} m(A_N),$$

and neither $m(A_0) = 0$ nor $m(A_0) > 0$ is compatible with $m([0, 1]) = 1$.

EXERCISES

(Exercises 1 through 5 are intended to be attacked "from first principles.")

1. Prove that if a set is in $\mathscr{I}$ (that is, a finite union of intervals) then it is a finite disjoint union of intervals. Prove also that if $I_1, I_2 \in \mathscr{I}$, then $I_1 \setminus I_2 \in \mathscr{I}$.

2. Prove that if I is a finite disjoint union of intervals, $I = \bigcup I_r$, then $m(I) = \sum m(I_r)$. Deduce that if $I \in \mathscr{I}$ is written as a finite disjoint union in two ways, $I = \bigcup I_r = \bigcup I'_s$, then $\sum m(I_r) = \sum m(I'_s)$ $[= m(I)$, by definition].

3. Prove that if $I_1, I_2 \in \mathscr{I}$ and $I_1 \subset I_2$, then $m(I_1) \le m(I_2)$.

4. Prove that if $\{I_r\}$ is a finite disjoint collection of sets of $\mathscr{I}$, then $m(\bigcup I_r) = \sum m(I_r)$.

5. By considering suitable polygons (finite unions of rectangles, in this case), prove that the area of the plane region bounded by $x = 1$, $y = 0$, $y = x^2$ is $1/3$. Generalize to the case $y = x^k$, where $k > 0$ is not necessarily an integer.

6. Defining the content of a plane set A to be the common value (if it exists) of inf $m(Q)$ and sup $m(P)$, where P and Q are polygons and $P \subset A \subset Q$, show that each set of a countable class may have content, whereas their union does not.

7. Develop the theory of measure for the class $\mathscr{J}$, starting from the definition

$$m(J) = \inf \sum m(I_r),$$

over all representations $J = \bigcup I_r$ as a countable disjoint union of intervals. Where (if at all) does compactness come into play in this development?

8. Prove directly that if a (not necessarily bounded) set J is written in two ways as a countable disjoint union of bounded intervals

$$J = \bigcup I_r = \bigcup I'_s,$$

then $\sum m(I_r) = \sum m(I'_s)$.

9. Defining the measure of a not necessarily bounded set J to be the common value of the two sums in Exercise 8, develop directly the theory of inner and outer measure for sets which are not necessarily bounded.

10. Prove that if $J \in \mathscr{J}(X)$, $\epsilon > 0$ are given, there is a subset J_0 of J which is in $\mathscr{I}$ and such that $m(J_0) > m(J) - \epsilon$. Show that the intervals of which J_0 is made up may be taken as all open or all closed.

11. Prove that the subset E of the bounded interval X is measurable if and only if, given $\epsilon > 0$, there exist J_1, $J_2 \in \mathcal{J}$ with $J_1 \supset A$, $J_2 \supset X \setminus A$, and

$$m(J_1) + m(J_2) < m(X) + \epsilon.$$

Is the result true if the class $\mathcal{J}$ is replaced by the class $\mathcal{I}$?

12. Prove that if $E^{(k)}$ is measurable for all sufficiently large k, then E is measurable.

13. Prove that if $m^*(E) = m_*(E) < \infty$, then E is measurable and $m(E) = m^*(E) = m_*(E)$. Give an example of a nonmeasurable set with $m^*(E) = m_*(E)$.

14. If $A_1 \subset A_2 \subset A_3 \subset \cdots$, and $A = \bigcup A_r$, is it true that $m_*(A) = \lim m_*(A_r)$? Give a proof or a counter example.

15. Dualize Theorem 2.3h.

16. Let f_0 be the map of a subset of the unit interval I on to a subset of the unit square S, defined by

$$f_0(0 \cdot x_1 x_2 x_3 \cdots) = (0 \cdot x_1 x_3 x_5 \cdots, 0 \cdot x_2 x_4 x_6 \cdots)$$

when the decimals on the right neither terminate nor end with $\dot{9}$. Show that f_0 can be extended to become a 1-1 map of I on to S. Prove that if f is any such extension, neither f nor its inverse f^{-1} can be continuous.

17. With the notation of Exercise 16, show that E is a measurable subset of I if and only if $f(E)$ is a measurable subset of S, and that then $m(E) = m(f(E))$. [Consider first intervals like $(k10^{-r}, (k+1)10^{-r})$ in I and the analogous intervals in S.] Generalize from R^2 to R^n. (This result indicates a possible—although rather artificial—procedure by which measure in R^n may be made to depend on measure in R.)

18. Prove that the Borel sets in R^n can be characterized as the smallest σ ring containing all open sets, and also as the smallest σ ring containing all compact sets.

19. If $\{A\}$ is a σ ring in R^n, and f a function on R^p to R^n, show that $\{f^{-1}(A)\}$ is a σ ring in R^p. Deduce that if f is a real continuous function and E is a Borel set in R, then $f^{-1}(E)$ is also a Borel set. If g is a function on R^n to R^q, is $\{g(A)\}$ a σ ring?

20. If $\{A\}$, $\{B\}$ are σ rings, show that $\{A \times B\}$ is not in general a σ ring. Show that if $\{A\}$, $\{B\}$ are each equal to the Borel sets in R, then the smallest σ ring containing $\{A \times B\}$ consists of the Borel sets in R^2. What about the corresponding result for Lebesgue-measurable sets?

21. Give an example of a continuous function f on $[0, 1]$ to itself, and a measurable subset E of $[0, 1]$ such that $f(E)$ is not measurable.

[3]

The Integral I

3.1. Definition

The elements of the theory of Lebesgue measure having been developed, the integral can be defined in various ways. One obvious—and attractive—method is to adopt the "area under the curve $y = f(x)$" idea of elementary calculus, and define the integral of a function to be the measure of its ordinate set. (See Sec. 3.5 for precise definitions and a discussion.) This approach leads to minor technical inconveniences in some places, and we prefer another (equivalent) definition.

Let E be a measurable set in R^n; by a *dissection* of E is meant a countable collection $\mathscr{E} = \{E_1, E_2, \cdots\}$ of disjoint measurable sets whose union is E. The dissection $\mathscr{E}'$ is a *refinement* of $\mathscr{E}$ if each set in $\mathscr{E}'$ is a subset of some set in $\mathscr{E}$. Given two dissections $\mathscr{E} = \{E_1, E_2, \cdots\}$ and $\mathscr{E}' = \{E_1', E_2', \cdots\}$, their *common refinement* $\mathscr{E} \vee \mathscr{E}'$ is the collection of all sets of the form $E_p \cap E_q'$; this is obviously again a dissection of E.

Let f be an extended-real function on E (that is, possibly taking the values $+\infty$ or $-\infty$), and let $\mathscr{E}$ be a dissection of E. For each relevant integer r, let

$$B_r(f) = \sup_{x \in E_r} f(x), \qquad b_r(f) = \inf_{x \in E_r} f(x),$$

$$h_r(f) = \sup_{x \in E_r} |f(x)| = \max\left(|b_r(f)|, |B_r(f)|\right).$$

Adopt the convention that if $h_r(f) = \infty$ and $m(E_r) = 0$, or if $h_r(f) = 0$ and $m(E_r) = \infty$, then the product $h_r(f)\, m(E_r)$ is to be taken as zero. A dissection $\mathscr{E}$ such that

$$\sum h_r(f)\, m(E_r) < \infty$$

will be called *admissible* (for f over E). The set of admissible dissections will be denoted by $\mathfrak{E}(f, E)$ or simply by $\mathfrak{E}$; it may of course be empty for given f, E. We assume in what follows that $\mathfrak{E}$ is not empty.

For $\mathscr{E} \in \mathfrak{E}$ the sums

$$S_\mathscr{E}(f) = \sum B_r(f)\, m(E_r), \qquad s_\mathscr{E}(f) = \sum b_r(f)\, m(E_r)$$

are defined without ambiguity, since the series, if infinite, are absolutely convergent. They are, respectively, the *upper* and *lower approximating sums* for

f over E corresponding to $\mathscr{E}$. If the function f is clear from the context, we may write B_r, $S_{\mathscr{E}}$ and so on, instead of $B_r(f)$, $S_{\mathscr{E}}(f)$; if the dissection in question is clearly understood, we may omit the suffix and write S for $S_{\mathscr{E}}$ (and S' for $S_{\mathscr{E}'}$, etc.).

LEMMA 3.1a. *If $\mathscr{E}$ is admissible for f over E, so is any refinement $\mathscr{E}'$ of $\mathscr{E}$; and $S' \le S$, $s' \ge s$.*

Proof. Denote by E'_{p1}, E'_{p2}, $\cdots$ the sets of $\mathscr{E}'$ which are contained in the set E_p of $\mathscr{E}$. It is trivial that $B'_{pq} \le B_p$, $b'_{pq} \ge b_p$, $h'_{pq} \le h_p$ for all p, q. Since $\sum_q m(E'_{pq}) = m(E_p)$, by Theorem 2.3g, it follows that

$$\sum_q B'_{pq} m(E'_{pq}) \le B_p m(E_p), \qquad \sum_q b'_{pq} m(E'_{pq}) \ge b_p m(E_p),$$

$$\sum_q h'_{pq} m(E'_{pq}) \le h_p m(E_p),$$

and the lemma now follows on summing over p.

The *upper integral* $\int_E^* f$ of f over E (with respect to Lebesgue measure) and the *lower integral* $\int_{*E} f$ are now defined by

$$\int_E^* f = \inf_{\mathscr{E} \in \mathfrak{E}} S_{\mathscr{E}}(f), \qquad \int_{*E} f = \sup_{\mathscr{E} \in \mathfrak{E}} s_{\mathscr{E}}(f).$$

(If the class $\mathfrak{E}$ is empty, it may still be possible in some cases to attach reasonable meanings to "upper integral" and "lower integral," but it is not necessary to go into this here.) If the set E is clearly implied by the context, we may write simply $\int^* f$, etc.; if, on the other hand, confusion is possible, a more explicit notation such as $\int_E^* f \, dm$ or $\int_E^* f(x) \, dx$ may be preferable. Note that under the restriction that $\mathfrak{E} \ne \varnothing$, the upper and lower integrals are both finite.

LEMMA 3.1b. *For any f, $\int^* f \ge \int_* f$.*

Proof. Let $\mathscr{E}$, $\mathscr{E}'$ be any two admissible dissections, and let $\mathscr{E}'' = \mathscr{E} \vee \mathscr{E}'$ be their common refinement. Then, by Lemma 3.1a,

$$s' \le s'' \le S'' \le S,$$

so that sup $s' \le S$ for all $\mathscr{E}$, $\mathscr{E}'$, and so sup $s' \le$ inf S.

There is no loss of generality in supposing that the dissections which figure in the definitions of the upper and lower integrals consist of bounded sets, for any set in R^n is a countable union of disjoint bounded sets, and so any dissection has a refinement consisting of bounded sets.

The function f is *integrable* (in the sense of Lebesgue, Lebesgue integrable) over the set E if the upper and lower integrals of f over E are equal, and their common value is defined to be the *integral* (with respect to Lebesgue measure) of f over E, written $\int_E f$ (or $\int f$, $\int_E f(x)\,dx$, etc.).

It is immediate that any function is integrable over a set of measure zero, its integral being zero. Likewise, the function which is identically zero is integrable over any set, and its integral is zero.

LEMMA 3.1c. *A necessary and sufficient condition that f should be integrable over E is that for any $\epsilon > 0$ there should be an admissible dissection $\mathscr{E}$ of E with $S_{\mathscr{E}} - s_{\mathscr{E}} < \epsilon$.*

Proof. If there is such a dissection, f is clearly integrable. Suppose f integrable; there are dissections $\mathscr{E}'$ and $\mathscr{E}''$ of E such that $S' - \int f < \epsilon/2$, $\int f - s'' < \epsilon/2$. Let $\mathscr{E} = \mathscr{E}' \vee \mathscr{E}''$; by Lemma 3.1a, $S \leq S'$ and $s'' \geq s$, whence $S - s < \epsilon$.

As a special case of the above result, if there is an admissible dissection $\mathscr{E}$ such that $S_{\mathscr{E}} = s_{\mathscr{E}}$, then f is integrable over E and $\int f = S_{\mathscr{E}} = s_{\mathscr{E}}$. This condition is satisfied in the case of a *generalized step function f*; that is, a function which takes only countably many values $f_1, f_2, \cdots$, and for which each set $E_r = \{x : f(x) = f_r\}$ is measurable. If the condition

$$\sum_r |f_r|\, m(E_r) < \infty$$

is satisfied, then the dissection $\mathscr{E} = \{E_1, E_2, \cdots\}$ is admissible and $S_{\mathscr{E}} = s_{\mathscr{E}} = \sum f_r m(E_r)$, so that f is integrable over E $(= \bigcup E_r)$ and

$$\int_E f = \sum f_r m(E_r).$$

In particular, let $f = \chi_E$, the characteristic function of a measurable set E (Sec. 1.5). Then χ_E is integrable (over any measurable set containing E) if and only if $m(E) < \infty$; and then

$$\int \chi_E = m(E).$$

Any constant function c is integrable over any set E of finite measure, and its integral has the value $cm(E)$.

Let E and E' be measurable, with $E \subset E'$. Let f be a function on E, and let g be the function defined on E' by

$$g(x) = f(x) \quad \text{if } x \in E$$
$$= 0 \qquad \text{if } x \in E' \setminus E.$$

LEMMA 3.1d. *The function f is integrable over E if and only if g is integrable over E', and, if so,*

$$\int_E f = \int_{E'} g.$$

Proof. Given any admissible dissection $\{E_1, E_2, \cdots\}$ of E, the dissection $\{E' \setminus E, E_1, E_2, \cdots\}$ of E' is also admissible and gives the same upper sum and the same lower sum. Conversely, given the dissection $\{E_1', E_2', \cdots\}$ of E', the dissection $\{E \cap E_1', E \cap E_2', \cdots\}$ gives the same sums. The required result follows at once.

One consequence of Lemma 3.1d is that all integrals over subsets of R^n may be regarded, if desired, as integrals over the whole of R^n. The integral of f over E is equal to the integral over R^n of the function equal to f in E and zero outside E. This point of view is sometimes useful.

The restriction that the set E over which a function is integrated should be measurable is not a serious one. Let A be a set which is not necessarily measurable, and let E be any measurable set containing A. If f is defined on A, let g be the function defined on E which is equal to f in A and zero outside A. Then we may say that f is integrable over A if g is integrable over E; and, if so, we may define $\int_A f = \int_E g$. By Lemma 3.1d, this is independent of E, and reduces to the usual definition if A is measurable. Also it can be shown (as an immediate corollary of Theorem 3.3e, below) that if f is integrable over A, then the subset of A on which f takes nonzero values must be measurable. The integral of f over A is equal to its integral over this subset (again by Lemma 3.1d). Since any integral over a nonmeasurable set can easily be reduced to one over a measurable set, we shall consider only the latter case.

It follows at once from the definition of integrability that if f is integrable over E (or, indeed, if the upper and lower integrals of f over E exist), then f is finite almost everywhere in E. For if not, then for any $\mathscr{E}$ at least one term in the series $\sum h_r m(E_r)$ would be infinite, and there would be no admissible dissections. The condition that f should be almost everywhere finite is, of course, not sufficient for integrability.

3.2. Elementary Properties

It is assumed throughout that all sets and functions are subject to sufficient restrictions for the assertions about them to make sense. For example, "if $f \geq g$ then $\int_E f \geq \int_E g$" should be interpreted as "if f and g are integrable over E, and $f \geq g$, then $\int_E f \geq \int_E g$."

THEOREM 3.2a. If $f = g$ p.p., then $\int^* f = \int^* g$, $\int_* f = \int_* g$, and $\int f = \int g$.

Proof. Let E_0 be the subset of E on which f and g are not equal. Given any dissection $\mathscr{E}$ of E, $\mathscr{E} = \{E_1, E_2, \cdots\}$, let $\mathscr{E}'$ be the dissection $\{E_0, E_1 \cap (E \setminus E_0), E_2 \cap (E \setminus E_0), \cdots\}$. Then it is clear that

$$S'(f) \leq S(g), \qquad S'(g) \leq S(f),$$

so that the upper integrals of f and g are equal. This is true also for the lower integrals, and the integrals (if the functions are integrable).

It follows that in integration theory it is sufficient if relations of equality or inequality (such as those of the next theorem) hold almost everywhere, and they should be taken in this sense unless the contrary is indicated. Indeed, functions need only be defined almost everywhere on the sets over which we wish to integrate. Let f be defined p.p. in E, and let g, h be functions defined on E and equal to f in its domain of definition. Then g and h are both integrable, with $\int_E g = \int_E h$, or both are nonintegrable; we may define f to be integrable over E if there is an integrable extension g defined on E, and write $\int_E f = \int_E g$.

THEOREM 3.2b. *If* $f \geq g$, *then* $\int^* f \geq \int^* g$, $\int_* f \geq \int_* g$, $\int f \geq \int g$.

Proof. This is obvious, since for any given dissection $S(f) \geq S(g)$ and $s(f) \geq s(g)$.

THEOREM 3.2c. *If* f *and* g *are integrable and* c *is a constant, then* cf *and* $f + g$ *are integrable, and*

$$\int cf = c \int f, \qquad \int (f + g) = \int f + \int g.$$

Proof. Evidently a dissection which is admissible for f is also admissible for cf. If $c \geq 0$, then $S(cf) = cS(f)$ and $s(cf) = cs(f)$; if $c < 0$, then $S(cf) = cs(f)$ and $s(cf) = cS(f)$. In either case the integrability of cf and the relation $\int cf = c \int f$ follow at once.

Let $\mathscr{E}'$, $\mathscr{E}''$ be admissible for f, g respectively; then $\mathscr{E} = \mathscr{E}' \vee \mathscr{E}''$ is admissible for both f and g. Since

$$B_r(f + g) \leq B_r(f) + B_r(g), \qquad b_r(f + g) \geq b_r(f) + b_r(g),$$

$$h_r(f + g) \leq h_r(f) + h_r(g),$$

it follows that $\mathscr{E}$ is admissible for $f + g$ and that

$$s(f) + s(g) \leq s(f + g) \leq S(f + g) \leq S(f) + S(g).$$

Given $\epsilon > 0$ we can take $\mathscr{E}'$, $\mathscr{E}''$ so that $S'(f) - s'(f) < \epsilon/2$, $S''(g) - s''(g) < \epsilon/2$; and so $S(f) - s(f) < \epsilon/2$, $S(g) - s(g) < \epsilon/2$. It follows that $S(f + g) - s(f + g) < \epsilon$; hence, by Lemma 3.1c, $f + g$ is integrable. The equation

$$\int (f + g) = \int f + \int g$$

is now obvious.

COROLLARY. *If* $f_1, f_2, \cdots, f_k$ *are integrable, and* $c_1, c_2, \cdots, c_k$ *are constants, then* $c_1 f_1 + c_2 f_2 + \cdots + c_k f_k$ *is integrable, and*

$$\int (c_1 f_1 + \cdots + c_k f_k) = c_1 \int f_1 + \cdots + c_k \int f_k.$$

The result of Theorem 3.2c may be expressed by saying that integration is a *linear operation*.

LEMMA 3.2d. If $E = E_1 \cup \cdots \cup E_k$, where the sets E_r are disjoint, then

$$\int_E f = \int_{E_1} f + \cdots + \int_{E_k} f.$$

Proof. Write χ_r for the characteristic function of E_r. Then, by Lemma 3.1d,

$$\int_E \chi_r f = \int_{E_r} f;$$

and the result now follows from Theorem 3.2c, Corollary since $\chi_1 f + \cdots + \chi_k f = f$.

It is also possible to give a direct proof of the above lemma, without using Theorem 3.2c.

For any real function f the *positive part* f^+ of f is defined by

$$\begin{aligned} f^+(x) &= f(x) \quad \text{if } f(x) \geq 0, \\ &= 0 \qquad \text{if } f(x) < 0. \end{aligned}$$

The *negative part* f^- of f is defined by $f = f^+ - f^-$; f^- is thus a nonnegative function, equal to 0 if $f \geq 0$, and equal to $-f$ if $f < 0$. It is clear that $|f| = f^+ + f^-$.

THEOREM 3.2e. If f is integrable, so are f^+, f^- and $|f|$; and $\int |f| \geq |\int f|$.

Proof. Let $\mathscr{E}$ be an admissible dissection of E, for f. Since for each r we have $h_r(f^+) \leq h_r(f)$, $\mathscr{E}$ is admissible also for f^+. And $B_r(f^+) - b_r(f^+) \leq B_r(f) - b_r(f)$, so that

$$S(f^+) - s(f^+) \leq S(f) - s(f),$$

and the integrability of f^+ follows by Lemma 3.1c. The integrability of f^- $(= f^+ - f)$ and of $|f|$ are now obvious, by Theorem 3.2c.

Finally we have (using Theorem 3.2c again)

$$\left| \int f \right| = \left| \int (f^+ - f^-) \right| = \left| \int f^+ - \int f^- \right| \leq \int f^+ + \int f^- = \int (f^+ + f^-) = \int |f|.$$

Let f and g be integrable functions; in view of the relations

$$f \vee g = (f + g + |f - g|)/2, \qquad f \wedge g = (f + g - |f - g|)/2$$

for the upper and lower envelopes of f and g (defined in Sec. 1.5), it is clear that $f \vee g$ and $f \wedge g$ are also integrable. The extension to any finite number of integrable functions is immediate.

LEMMA 3.2f. If E is measurable, and A is subset of E such that $m^*(A) < \infty$, then

$$\int_E^* \chi_A = m^*(A), \qquad \int_{*E} \chi_A = m_*(A).$$

Proof. Since A is of finite outer measure, there is a measurable set H of finite measure containing it (a set in $\mathcal{J}$, for instance). There is thus an admissible dissection for χ_A over E (one such dissection consists of H and its complement). Now

$$\int_E^* \chi_A \leq \int_E^* \chi_H = \int_E \chi_H = m(H),$$

and $m^*(A) = \inf m(J)$ for $J \in \mathcal{J}$, $J \supset A$ (actually $m^*(A) = \inf m(H)$ for all measurable $H \supset A$) so that $\int_E^* \chi_A \leq m^*(A)$. On the other hand, if $\mathscr{E} = \{E_1, E_2, \cdots\}$ is any admissible dissection of E, we have $S_{\mathscr{E}}(\chi_A) = \Sigma' m(E_r)$, the sum being over all r such that $A \cap E_r \neq \varnothing$. If the union of all such sets E_r is denoted by E', then

$$S_{\mathscr{E}}(\chi_A) = m(E') = m^*(E') \geq m^*(A),$$

so that $m^*(A) \leq \inf S_{\mathscr{E}}(\chi_A) = \int_E^* \chi_A$. This proves the result stated for upper integrals; the proof for lower integrals is dual.

LEMMA 3.2g. *If f is integrable over E and E' is a measurable subset of E, then f is integrable over E'.*

Proof. If $\mathscr{E} = \{E_1, E_2, \cdots\}$ is admissible for f over E, and for each r, $E'_r = E_r \cap E'$, then evidently $\mathscr{E}' = \{E'_1, E'_2, \cdots\}$ is admissible for f over E'. Also $B'_r - b'_r \leq B_r - b_r$ for all r, so that $S' - s' \leq S - s$; the integrability of f over E' now follows from the criterion of Lemma 3.1c.

THEOREM 3.2h. *If $f \geq 0$, a necessary and sufficient condition that $\int f = 0$ is that $f = 0$ p.p.*

Proof. It is, of course, trivial that if $f = 0$ p.p., then $\int f = 0$. In order to prove the converse, suppose first that E is of finite measure. For each positive integer k, let A_k be the subset of E on which $f \geq k^{-1}$; then A_k is of finite outer measure, and

$$0 = \int kf = \int^* kf \geq \int^* \chi_{A_k} = m^*(A_k)$$

(by Lemma 3.2f), and so A_k is of measure zero. The union of all the sets A_k is therefore of measure zero; but this is the set on which f is nonzero.

If E is not of finite measure, let E' be a subset of E of finite measure; let g be equal to f in E' and zero outside E'. Then f is integrable over E' (Lemma 3.2g) and so g is integrable over E (Lemma 3.1d), and, since $f \geq g$ on E,

$$0 = \int_E f \geq \int_E g = \int_{E'} f \geq 0,$$

so that $\int_{E'} f = 0$, and so $f = 0$ p.p. in E'. Since E is a countable union of subsets of finite measure, $f = 0$ p.p. in E also.

THEOREM 3.2i. If f is integrable over E, then $\int_F f \to 0$ as $m(F) \to 0$, uniformly for subsets F of E.

Proof. Let $\mathscr{E}$ be a fixed admissible dissection of E. Given $\epsilon > 0$, there is an N such that $\sum_{r=N+1}^{\infty} h_r m(E_r) < \epsilon/2$. Write $h = \sup' h_r$, the supremum being over all r with $1 \leq r \leq N$ such that h_r is finite (if $h_r = \infty$, then $m(E_r)$ must be zero and $h_r m(E_r) = 0$). Then if F is any subset of E such that $m(F) < \epsilon/2h$, we have $\mid \int_F f \mid < \epsilon$.

For, let $\mathscr{E}'$ be the dissection of F induced by $\mathscr{E}$; then $\mid \int_F f \mid \leq \int_F \mid f \mid$, and

$$\int_F \mid f \mid \leq \sum_{r=1}^{\infty} h'_r m(E'_r)$$

$$\leq h \sum_{r=1}^{N} m(E'_r) + \sum_{r=N+1}^{\infty} h'_r m(E'_r)$$

$$\leq hm(F) + \sum_{r=N+1}^{\infty} h_r m(E_r)$$

$$< \epsilon/2 + \epsilon/2 = \epsilon.$$

(It has been assumed in the above that the dissection $\mathscr{E}$ is not finite; this, of course, involves no loss of generality.)

COROLLARY. If $E_1 \subset E_2 \subset E_3 \cdots$, $E = \bigcup E_r$ is bounded and f is integrable over E, then

$$\int_E f = \lim \int_{E_r} f.$$

Proof. This follows at once in view of Theorem 2.3h.

3.3. Measurable Functions

Roughly speaking, a function is integrable if its behavior is not too irregular, and if the values which it takes are not too large too often. The second requirement is equivalent to the existence of an admissible dissection; the first to the equality of the upper and lower integrals. We now introduce the notion of *measurability*, which will be shown to provide precisely the condition required for integrability, given that the function is not too large. In many cases it is easier to examine the measurability of a function than to investigate its upper and lower integrals directly.

Let E be a measurable set and f an extended-real function on E. We say that f is *measurable in* E if for each real a the subset of $E\{x: f(x) > a\}$ $(= f^{-1}(]a, \infty]))$ is measurable. (This implies at once that the sets $\{x: f(x) = +\infty\}$,

$\{x: f(x) = -\infty\}$ are measurable.) Usually, the set E will not be mentioned explicitly; this should cause no confusion.

There are several minor variations in the basic definition, all equivalent to the version just given. Since the sets $\{x: f(x) > a\}$ and $\{x: f(x) \leq a\}$ are complementary, if either is measurable, so is the other, and thus the second may replace the first in the definition of measurability. Also, if each set of the form $\{x: f(x) \geq a\}$ is measurable, so is $\bigcup_{r=1}^{\infty} \{x: f(x) \geq a + r^{-1}\} = \{x: f(x) > a\}$. If this is measurable for each a, then each set $\{x: f(x) \leq a\}$ is measurable, and so $\{x: f(x) < a\} = \bigcup_{r=1}^{\infty} \{x: f(x) \leq a - r^{-1}\}$ is measurable. It follows that $\{x: f(x) > a\}$ could be replaced in the definition by either $\{x: f(x) < a\}$ or $\{x: f(x) \geq a\}$.

It is immediate that a necessary and sufficient condition for measurability is that $\{x: a \leq f(x) \leq b\}$ should be measurable for all a, b (including the cases $a = -\infty$, $b = +\infty$), for any set of this form can be written as the intersection of two sets: $\{x: f(x) \geq a\} \cap \{x: f(x) \leq b\}$; if f is measurable, each of these is measurable, and so is $\{x: a \leq f(x) \leq b\}$. Conversely, any set of the form occurring in the definition can easily be expressed in terms of sets of the form $\{x: a \leq f(x) \leq b\}$.

Another immediate consequence of the definition is that if two functions are equal almost everywhere, and one is measurable, so is the other, for if $f = g$ p.p., then $\{x: f(x) > a\} = (\{x: g(x) > a\} \cup E_1) \setminus E_2$, where E_1 and E_2 are subsets of the set on which $f \neq g$, hence of measure zero. So if one set is measurable, so is the other, and this is true also for the sets on which the functions take infinite values. As in the case of integrability, we shall regard a function as adequately defined (for a discussion of its measurability) if it is merely defined almost everywhere in E, instead of on the whole of E. Then if any extension to the whole of E is measurable, so is any other, and we regard the original function as measurable on E.

The sum of two extended-real functions is—from the point of view just explained—adequately defined provided that the set on which one function takes the value $+\infty$ and the other takes the value $-\infty$ is of measure zero. Similarly, the product is adequately defined if the set on which one function is zero and the other is infinite is of measure zero. These conditions are certainly satisfied if the functions are finite almost everywhere (which is the case of primary interest here). In what follows it is assumed that the functions are so restricted that the combination in question is adequately defined, and similarly for functions of f such as $f^{1/k}$ (where, if k is an even integer, f must be nonnegative).

LEMMA 3.3a. If f and g are measurable functions and c is a constant, then $f + c$, cf, $f + g$, f^2, $f^{1/2}$, fg, $f \vee g$, $f \wedge g$, f^+, f^-, $|f|$ are measurable.

Proof. The cases $f + c$ and cf are trivial. For $f + g$,

$$\{x: f(x) + g(x) > a\} = \bigcup \{x: f(x) > t\} \cap \{x: g(x) > a - t\},$$

the union being over all *rational* t; the set on the left is a countable union of measurable sets and is therefore measurable.

If $a < 0$ the set $\{x: f^2(x) > a\}$ is the whole of E, therefore measurable: if $a \geq 0$, then

$$\{x: f^2(x) > a\} = \{x: f(x) > a^{1/2}\} \cup \{x: f(x) < -a^{1/2}\}$$

and again the set is measurable. The set on which f^2 is infinite is the union of the sets on which f is $+\infty$ or $-\infty$, therefore measurable. Similar reasoning applies to $f^{1/2}$. The measurability of fg follows from the identity

$$fg = ((f + g)^2 - f^2 - g^2)/2.$$

Since $\{x: f^+(x) > a\} = \{x: f(x) > a\}$ if $a \geq 0$, $= E$ if $a < 0$, f^+ is measurable if f is measurable; and similarly for f^- (or, by noting that $f^- = f^+ - f$). The measurability of $|f|$ now follows at once, as does the measurability of $f \vee g$ and $f \wedge g$, in view of the relations $|f| = f^+ + f^-$, $f \vee g = (f + g + |f - g|)/2$, $f \wedge g = (f + g - |f - g|)/2$.

COROLLARY. The sum, product, upper envelope, and lower envelope of any finite number of measurable functions are measurable.

There are various general classes of functions which can be shown to be measurable. For instance if a function of a single real variable is monotonic (increasing if $x \geq y$ implies $f(x) \geq f(y)$, decreasing if $x \geq y$ implies $f(x) \leq f(y)$), then it is measurable. Take the increasing case; suppose the domain of definition of the function is E. If $b = \inf x$ for $f(x) > a$, $x \in E$, then $\{x: f(x) > a\}$ is either $E \cap]b, \infty]$ or $E \cap [b, \infty]$ and is measurable in either case; hence f is measurable. Another important case is described in the following theorem:

THEOREM 3.3b. If a real function is continuous, it is measurable.

Proof. Suppose f is defined on E and is continuous in E. Then $f^{-1}(]a, \infty[)$ is open relative to E (Theorem 1.5a), hence of the form $E \cap F$, where F is open in R^n (Lemma 1.3e), hence the intersection of two measurable sets, and therefore measurable. In the case of a continuous real function, the question of infinite values does not arise.

One may define continuity also for extended-real functions, and then the statement and proof of the above theorem remain valid, with only slight modifications. The theorem as given, however, is adequate for our requirements, since we are concerned with functions which are almost everywhere finite; if such a function is continuous at each point where it is finite, then it is measurable.

Although Lemma 3.3a shows that measurable functions can be combined in various ways, to produce a measurable function, their behavior in this respect is not as satisfactory as that of continuous functions. One defect is that, if f

is a measurable function on R, and g is measurable, it is not in general true that the composite function $f(g(\cdot))$ is measurable. For example, take Cantor's set T and the functions defined at the end of Sec. 1.2. If A is a nonmeasurable subset of $[0, 1]$, the composite function $\phi(\cdot) = \chi_A(h(\cdot))$ is nevertheless measurable, as $\{x\colon \phi(x) > a\}$ is either a subset of T, or the complement of a subset of T with respect to $[0, 1]$, according as $0 \notin A$ or $0 \in A$. In either case the set is measurable. The function f defined in Sec. 1.2, being an increasing function, is measurable. The composite function $\phi(f(\cdot))$, however, is simply $\chi_A(\cdot)$ and is not measurable.

One can establish various results, giving conditions under which the composite function is measurable; perhaps the most useful is the following.

LEMMA 3.3c. *If f is a continuous function on R, and g is measurable, then $f(g(\cdot))$ is measurable.*

Proof. Since f is continuous, $f^{-1}(]a, \infty[)$ is open and is therefore a countable union of intervals I_r (Theorem 2.2g). Also, for each r the set $\{x\colon g(x) \in I_r\}$ is measurable, being the intersection of two measurable sets of the type figuring in the definition. And since

$$\{x\colon f(g(x)) > a\} = \{x\colon g(x) \in f^{-1}(]a, \infty[)\}$$
$$= \bigcup \{x\colon g(x) \in I_r\},$$

the set $\{x\colon f(g(x)) > a\}$ is measurable, and so the function $f(g(\cdot))$ is measurable.

The above lemma applies, strictly speaking, only to the case where g is everywhere finite, but it extends immediately to the case where g is finite almost everywhere. As in the case of Theorem 3.3b, one could formulate the result in terms of continuous extended-real functions defined on the extended-real line, with no essential change; but this is not necessary for our purpose. In many instances it is possible to use the lemma just given, with trivial complements, to examine functions which are not almost everywhere finite; for example, one deduces at once that if f is measurable, so is $f/(1 + f^2)$, whether or not f is finite p.p.

One useful feature of measurable functions is that various countably infinite processes applied to such functions yield a function which is again measurable. In particular:

LEMMA 3.3d. *If $f_1, f_2, \cdots$ are measurable, so are $\bigvee_{r=1}^{\infty} f_r$, $\bigwedge_{r=1}^{\infty} f_r$, $\limsup f_r$, $\liminf f_r$, and $\lim f_r$ (if the sequence is convergent).*

Proof. Write g for the upper envelope $\bigvee_{r=1}^{\infty} f_r$. Then

$$\{x\colon g(x) > a\} = \bigcup_{r=1}^{\infty} \{x\colon f_r(x) > a\},$$

and this set is measurable. The upper envelope is therefore measurable, and a similar argument applies to the lower envelope.

Since $\limsup f_r = \bigwedge_{s=1}^{\infty} (\bigvee_{r=s}^{\infty} f_r)$, and $\liminf f_r = \bigvee_{s=1}^{\infty} (\bigwedge_{r=s}^{\infty} f_r)$, the upper and lower limits are measurable, by the result just proved. Finally, if the sequence is convergent, its limit is the common value of $\limsup f_r$ and $\liminf f_r$, and hence is also measurable.

The next theorem establishes the connection between integrability and measurability and, indeed, provides the major justification for introducing the latter idea. It is convenient to define a function f to be *dominated* in E (by g) if there is an integrable function g such that $|f| \le g$ throughout E.

THEOREM 3.3e. A necessary and sufficient condition that f should be integrable over E is that it should be measurable and dominated in E.

Proof. Let f be measurable and dominated by g; take an admissible dissection for g. We may assume, without loss of generality, that all sets in this dissection are of finite measure, and also (for convenience) that there are infinitely many of them. Given $\epsilon > 0$, choose P so that

$$\sum_{r=P+1}^{\infty} h_r(g) \, m(E_r) < \epsilon/3.$$

Writing $E_0 = E_1 \cup \cdots \cup E_P$, choose N so that $N^{-1} m(E_0) < \epsilon/3$. Now consider the dissection $\mathscr{E}'$ obtained by taking the intersections of the sets of the original dissection with all sets of the form

$$\{x : p/N \le f(x) < (p+1)/N\}, \qquad \cdot$$

where p is an integer, and the sets on which f is $+\infty$ or $-\infty$. This dissection is evidently admissible for f, and

$$S' - s' = \left(\sum_1 + \sum_2 \right) (B_r(f) - b_r(f)) \, m(E_r'),$$

where the first sum is over all r such that $E_r' \subset E_0$, and the second over all r such that $E_r' \subset E \setminus E_0$. Now, obviously, $\sum_1 \le N^{-1} m(E_0) < \epsilon/3$, and

$$\sum_2 \le \sum_2 (|B_r(f)| + |b_r(f)|) \, m(E_r')$$

$$\le 2 \sum_{p=P+1}^{\infty} h_p(g) \, m(E_p)$$

$$< 2\epsilon/3$$

so that $S' - s' < \epsilon$ and f is integrable.

Suppose that f is integrable; if $\{E_1, E_2, \cdots\}$ is an admissible dissection, the function g defined by

$$g(x) = h_r(f) \quad \text{if } x \in E_r$$

is evidently integrable ($\int g = \Sigma h_r m(E_r)$) and f is dominated by g.

It remains to show that if f is integrable it is measurable. For any dissection of E, define

$$\Phi(x) = B_r(f) \quad \text{if } x \in E_r,$$
$$\phi(x) = b_r(f) \quad \text{if } x \in E_r;$$

then Φ and ϕ are measurable, $\int \Phi = S$, $\int \phi = s$, and $\Phi \geq f \geq \phi$. Now take a sequence of dissections $\mathscr{E}_1$, $\mathscr{E}_2$, $\cdots$, each a refinement of the preceding one, such that

$$S_r \to \int f, \qquad s_r \to \int f.$$

The corresponding sequences Φ_r, ϕ_r are respectively decreasing and increasing; hence their limits Ψ, ψ exist and are measurable. Indeed, they are dominated (by $|\phi_1| + |\Phi_1|$, for example), hence integrable. Since evidently $\Psi \geq f \geq \psi$, we have $\int \Psi \geq \int f \geq \int \psi$. But also $\int \Psi \leq \int \Phi_r$ for all r, and $\int \Phi_r \to \int f$, so that $\int \Psi \leq \int f$; hence $\int \Psi = \int f$, and similarly $\int \psi = \int f$. We then have $\Psi - \psi \geq 0$, and $\int (\Psi - \psi) = 0$, so that $\Psi = \psi$ p.p., by Theorem 3.2h. Since $\Psi \geq f \geq \psi$, $f = \Psi$ p.p., and so f is measurable.

COROLLARY 1. If E is of finite measure and f is bounded, then f is integrable if and only if it is measurable.

COROLLARY 2. A bounded continuous function is integrable over any set of finite measure.

COROLLARY 3. If f is measurable, then f is integrable if and only if $|f|$ is integrable.

COROLLARY 4. If f is integrable and g is measurable and bounded, then fg is integrable.

3.4. Complex and Vector Functions

Integrability and measurability for such functions are defined in the obvious way in terms of real functions. We say that $f = g + ih$ is integrable, or measurable, if both g and h have the property in question. The vector function $f = (f_1, \cdots, f_p)$ is integrable or measurable if each of its component functions f_r is so. If f is integrable, its integral is $\int g + i \int h$ or $(\int f_1, \cdots, \int f_p)$, as the case may be.

Many theorems carry over at once from the real to the complex or vector case—for instance, the theorem that the sum of two integrable functions is integrable, and the integral of the sum is the sum of the integrals. Such results will be used freely when required, without comment in general. There is one result which we shall prove here, the extension of the inequality $|\int f| \le \int |f|$ (already proved for real functions) to the complex and vector situation. It is convenient to prove two results on real functions as a preliminary:

LEMMA 3.4a. If f and g are real measurable functions such that f^2 and g^2 are integrable, then fg is integrable, and

$$\left(\int fg\right)^2 \le \int f^2 \int g^2.$$

Proof. Since fg is measurable and dominated by $f^2 + g^2$, it is integrable. Hence for each real λ the function $(f + \lambda g)^2$ is integrable; since it is nonnegative its integral is nonnegative; that is,

$$\int f^2 + 2\lambda \int fg + \lambda^2 \int g^2 \ge 0$$

for all real λ; by a well-known elementary property of quadratics, this implies the inequality required.

(The inequality of Lemma 3.4a is usually called *Schwarz's inequality;* the names of other mathematicians—Cauchy in particular—are sometimes associated with it.)

LEMMA 3.4b. For any real integrable functions f, g,

$$\left(\int f\right)^2 + \left(\int g\right)^2 \le \left(\int (f^2 + g^2)^{1/2}\right)^2.$$

Proof. Let $F = (f^2 + g^2)^{1/4}$, and let $G = f(f^2 + g^2)^{-1/4}$ (if $f^2 + g^2 \ne 0$, and 0 if $f^2 + g^2 = 0$; $G = \pm \infty$ if $f = \pm \infty$). Then it is immediate that F and G are measurable; they are of integrable square, since F^2, G^2 are dominated by $|f| + |g|$, $|f|$, respectively. By Lemma 3.4a we have

$$\left(\int f\right)^2 = \left(\int FG\right)^2 \le \int F^2 \int G^2.$$

Similarly, if $H = g(f^2 + g^2)^{-1/4}$, then

$$\left(\int g\right)^2 = \left(\int FH\right)^2 \le \int F^2 \int H^2.$$

Hence $(\int f)^2 + (\int g)^2 \le \int F^2 \int (G^2 + H^2)$; but $G^2 + H^2 = F^2 = (f^2 + g^2)^{1/2}$, and the required result follows.

COROLLARY. If $f_1, \cdots, f_p$ are real integrable functions, then $(\int f_1)^2 + \cdots + (\int f_p)^2 \le (\int (f_1^2 + \cdots + f_p^2)^{1/2})^2$.

THEOREM 3.4c. If f is an integrable complex function, then $|f|$ is integrable and

$$\left| \int f \right| \le \int |f|;$$

if f is an integrable vector function, then $\|f\|$ is integrable and

$$\left\| \int f \right\| \le \int \|f\|.$$

Proof. This is simply a restatement in other terms of Lemma 3.4b and its corollary.

3.5. Other Definitions of the Integral

In this section all functions are again supposed to be real, or extended real.

One possible alternative approach to the integral (due to Lebesgue) is as follows: Suppose, for simplicity, that the function f is bounded, $a < f(x) < b$, and that the set E is of finite measure. Given $\epsilon > 0$, choose N so that $N^{-1}m(E) < \epsilon$. Let E_r be the set

$$\{x\colon a + (r - 1) N^{-1} \le f(x) < a + rN^{-1}\};$$

if f is measurable, then each E_r is measurable. If S, s are the upper and lower approximating sums (finite sums, here) corresponding to the dissection $\{E_1, E_2, \cdots\}$, then it is immediate that $S - s \le N^{-1} \sum m(E_r) = N^{-1}m(E) < \epsilon$, so that f is integrable by the definition of Sec. 3.1. With measurability assumed, the integral may be defined as the common value of $\inf S$ and $\sup s$ for dissections of the type just described.

Another definition, already mentioned in Sec. 3.1, is based on a certain subset of R^{n+1}. Suppose, for the time being, that f is a nonnegative function, E a measurable subset of R^n. The subset $\Omega(f, E)$ of R^{n+1}, defined by

$$\Omega(f, E) = \{x\colon (x_1, \cdots, x_n) \in E, 0 \le x_{n+1} < f(x_1, \cdots, x_n)\}$$

is the *ordinate set* of f over E. (We have preferred to write $x_{n+1} < f(x_1, \cdots, x_n)$ rather than $x_{n+1} \le f(x_1, \cdots, x_n)$ in the definition for convenience in handling extended-real functions. There is no essential difference.) Then one may define f to be integrable over E if $\Omega(f, E)$ is measurable and of finite measure; and

$$\int_E f = m(\Omega(f, E)).$$

It is possible to show that this is equivalent to the definition of Sec. 3.1. We follow the development given by Burkill in *The Lebesgue Integral*; first we establish a result which logically belongs to Sec. 2.3.

LEMMA 3.5a. If A is a subset of R^n, I an interval in R^p, then

$$m^*(A \times I) = m^*(A)\, m(I), \qquad m_*(A \times I) = m_*(A)\, m(I).$$

Proof. There is no loss of generality in assuming that the sets are bounded and that the interval I is closed. By Theorem 2.3i, if E and E' are measurable, then so is $E \times E'$ and $m(E \times E') = m(E)\, m(E')$. If $J \supset A$, then $J \times I \supset A \times I$, and so

$$m^*(A \times I) \leq m(J \times I) = m(J)\, m(I);$$

hence $m^*(A \times I) \leq \inf m(J)\, m(I) = m^*(A)\, m(I)$.

It remains to show that $m^*(A)\, m(I) \leq m^*(A \times I)$. Let $J' \supset A \times I$, with $m(J') < m^*(A \times I) + \epsilon/2$. If $J' = \bigcup I'_r$, a countable union of disjoint intervals, take for each r an open interval $I''_r \supset I'_r$, with $m(I''_r) < m(I'_r) + \epsilon 2^{-r-2}$. Then if $J'' = \bigcup I''_r$, $m(J'') \leq \sum m(I''_r) < \sum m(I'_r) + \epsilon/2 < m^*(A \times I) + \epsilon$.

For each $x \in A$, the section of J'' by x is a countable union of open intervals covering the section of $A \times I$ by x, that is, the interval I. Since I is bounded and closed, it is compact. The covering of I therefore has a finite refinement. The projection on R^n of each interval of J'' which corresponds to an interval of this refinement is an open interval containing x. The intersection of all the open intervals obtained in this way is again an open interval containing x. The union—for all x—of such intervals is an open set J containing A.

Evidently $J \times I \subset J''$, and so $m(J)\, m(I) \leq m(J'') < m^*(A \times I) + \epsilon$, whence $m^*(A)\, m(I) < m^*(A \times I) + \epsilon$; since ϵ is arbitrary, $m^*(A)\, m(I) \leq m^*(A \times I)$, and the result follows.

The corresponding result for inner measures is deduced at once, on taking complements.

LEMMA 3.5b. If $\Omega(f, E)$ is measurable, then f is a measurable function.

Proof. It is required to establish that the set

$$A(a) = \{x : f(x) > a\}$$

is measurable for each finite a. It may be assumed that $m(E)$ is finite. For any positive integer k, let I be the interval $[a, a + k^{-1}]$ of R. Then

$$A(a) \times I \supset \Omega \cap (R^n \times I) \supset A(a + k^{-1}) \times I,$$

and so, by Lemma 3.5a,

$$m_*(A(a))\, m(I) \geq m(\Omega \cap (R^n \times I)) \geq m^*(A(a + k^{-1}))\, m(I),$$

giving $m_*(A(a)) \geq m^*(A(a + k^{-1}))$ for any k.

Now, the sets $A(a + k^{-1})$ form an increasing sequence whose union is $A(a)$. By Theorem 2.3h, $m^*(A(a)) = \lim m^*(A(a + k^{-1}))$, and it follows that $m_*(A(a)) \geq m^*(A(a))$, showing that $A(a)$ is measurable.

THEOREM 3.5c. A necessary and sufficient condition that f should be integrable (in the sense of Sec. 3.1) over E is that $\Omega(f, E)$ should be measurable and of finite measure; if this is so, then

$$\int_E f = m(\Omega(f, E)).$$

Proof. Suppose f integrable: let $\{E_1, E_2, \cdots\}$ be an admissible dissection such that $S - s < \epsilon$. Let $I_r = [0, B_r[$, $I'_r = [0, b_r[$. Then if $F = \bigcup E_r \times I_r$, and $F' = \bigcup E_r \times I'_r$, it is clear that $F \supset \Omega \supset F'$, and so $m^*(\Omega) \leq m(F) = S$, $m_*(\Omega) \geq m(F') = s$. Hence $m^*(\Omega) \leq \inf S = \int f$ and $m_*(\Omega) \geq \sup s = \int f$, so that Ω is measurable and $m(\Omega) = \int f$.

Conversely, suppose that Ω is measurable and of finite measure; by Lemma 3.5b f is measurable, and it only remains to prove that it is dominated. Let $E_r = \{x: 2^r \leq f(x) < 2^{r+1}\}$ for any (positive or negative) integer r; let $E_{-\infty}$ be the set on which f is zero, and E_∞ the set on which it is infinite. Let I_r be $[0, 2^r]$, $I_{-\infty} = \{0\}$, $I_\infty = [0, \infty[$. Then

$$\bigcup E_r \times I_{r+1} \supset \Omega \supset \bigcup E_r \times I_r$$

(where the unions include $r = -\infty$, $r = \infty$) and so

$$\sum m(I_r)\, m(E_r) \leq m(\Omega).$$

But this implies that $\sum m(I_{r+1})\, m(E_r) = 2 \sum m(I_r)\, m(E_r)$ is finite. Hence, if the function g is defined by

$$\begin{aligned}
g(x) &= 0 && \text{if } x \in E_{-\infty}\\
&= 2^{r+1} && \text{if } x \in E_r\\
&= \infty && \text{if } x \in E_\infty,
\end{aligned}$$

then g is evidently integrable (its integral being $\sum m(I_{r+1})\, m(E_r)$) and f is dominated by g.

One slight disadvantage of defining the integral of a function to be the measure of its ordinate set is that the definition applies in the first place only to nonnegative functions. The general case, however, is easily treated; one has only to write $f = f^+ - f^-$, and consider f^+ and f^- separately. One very valuable feature of this definition is that theorems on measurable sets can frequently be translated into terms of integrals with very little trouble; we shall, on occasion, take advantage of this.

The procedure we adopted in Sec. 3.1 may be looked at from the following point of view: We have a class of functions—the generalized step functions— for which the integral has been defined. The integral can then be extended to those functions which admit arbitrarily close upper and lower approximations by functions of the class in question. The resulting functions have all the properties we may reasonably expect.

Now, the same sort of thing may be done, starting from any class of functions in which an "elementary integral" is defined. For example, we could take the class of step functions in the strict sense (functions taking only a finite set of values, each on a finite union of intervals). If the class from which we start is relatively complicated—such as the generalized step functions—then the class of "integrable" functions is relatively large and is adequate for a reasonable theory of integration. If, on the other hand, the initial class is simple, the resulting "integrable" functions may be inadequate, and the process must be repeated. It is indeed a characteristic of integration theories that, if one starts from the intervals, then a two-stage process is necessary to reach an adequate class of measurable sets or integrable functions.

If one sets up a theory of integration without explicitly developing the theory of measure, then the measure of a set can easily be defined in terms of the integral; a set is measurable and of finite measure if and only if its characteristic function is integrable, and its measure is the integral of the function.

EXERCISES

1. Prove that the function f is integrable if and only if, for any $\epsilon > 0$, there exist integrable functions g and h, with $g \geq f \geq h$, such that $\int g - \int h < \epsilon$. Show also that in this criterion g and h may be required to be generalized step functions, but not, in general, step functions in the strict sense.

2. Prove that if (a) $m(E) < \infty$, (b) f is bounded, then

$$\int_E^* f = \inf S_{\mathscr{E}}, \qquad \int_{*E} f = \sup s_{\mathscr{E}},$$

where the inf and sup are over all *finite* admissible dissections $\mathscr{E}$ of E. Show that if either (a) or (b) is relaxed, the result is false in general.

3. If $\mathscr{E} = \{E_r\}$ is an admissible dissection of E, and $X = \{x^r\}$ a countable subset of E such that $x^r \in E_r$ for all r, we define

$$\sigma(\mathscr{E}, X, f) = \sum f(x^r) \, m(E_r).$$

Prove that the (real or complex) function f is integrable over E if and only if, given $\epsilon > 0$, there is an admissible dissection $\mathscr{E}$ of E such that

$$| \, \sigma(\mathscr{E}, X, f) - \sigma(\mathscr{E}', X', f) \, | \, < \epsilon$$

whenever $\mathscr{E}'$ is a refinement of $\mathscr{E}$, for any X, X'. State and prove the corresponding result for vector functions.

4. If $\int_E f$ exists, we define, for $\delta > 0$, (with the notation of Exercise 3)

$$\phi(\delta) = \sup | \int_E f - \sigma(\mathscr{E}, X, f) \, |,$$

the supremum being over all $\mathscr{E}$ such that $m(E_r) \leq \delta$ for all r, and all X. Prove that $\phi(\delta) \to 0$ as $\delta \to 0$. If also

$$\psi(\delta) = \sup | \sigma(\mathscr{E}, X, f) - \sigma(\mathscr{E}', X', f) |,$$

over all $\mathscr{E}$, $\mathscr{E}'$ such that $m(E_r) \leq \delta$, $m(E_r') \leq \delta$ for all r, and all X, X', prove that f is integrable over E if and only if $\psi(\delta) \to 0$ as $\delta \to 0$.

5. Let $\{A_r\}$ be a "dissection" of E into a countable class of not necessarily measurable sets. Define, for real f,

$$S^* = \sum B_r m^*(A_r), \qquad s_* = \sum b_r m_*(A_r).$$

Is it true that f is integrable over E if and only if inf $S^* = \sup s_*$?

6. E is the subset of $[0, 1]$ consisting of all $x = 0 \cdot x_1 x_2 x_3 \cdots$ (in decimal notation) such that $x_r = x_{r+1} = \cdots = x_{2r} = 7$ for some r. Show that E is measurable. Prove that x^{-1} is integrable over E, but that x^{-2} is not. What happens if 7 is replaced by some other integer?

7. Give a direct proof of Lemma 3.2d.

8. Prove (with the notation of Sec. 1.5) that

$$\int f \wedge \int g \geq \int f \wedge g.$$

If the two sides are equal, what can be deduced about the relation between f and g?

9. Prove (a) directly, (b) using Theorem 3.5c, that if $\int_E f$ exists, and $\epsilon > 0$ is given, there is a bounded subset E' of E such that

$$| \int_E f - \int_{E'} f | < \epsilon.$$

(Compare Lemma 4.1c.)

10. Deduce from Exercise 9 that in the corollary to Theorem 3.2i the set E need not be bounded and need not be of finite measure.

11. If E is a countable disjoint union, $E = \bigcup E_r$, and $\int_E f$ exists, then $\int_E f = \sum \int_{E_r} f$.

12. Prove that f is a measurable function if and only if $f^{-1}(G)$ is a measurable set whenever G is open in R.

13. Prove that if a function is continuous p.p., then it is measurable. Give an example of a measurable function which is nowhere continuous.

14. Prove that if f is measurable and E is a Borel set in R, then $f^{-1}(E)$ is measurable.

15. Give an example of a measurable function f and a measurable set $E \subset R$ such that $f^{-1}(E)$ is not measurable. (A suitable subset of Cantor's set and one of the associated functions will serve.) What does this imply about the relation between Borel sets and Lebesgue-measurable sets?

16. We define a *Baire function* to be a function f such that $f^{-1}(G)$ is a Borel set whenever G is open. Prove that if f is a Baire function on R, and g is a real measurable function, then $f(g(\cdot))$ is measurable. If g is a Baire function, is $f(g(\cdot))$ necessarily a Baire function?

17. The extended-real function f is measurable, is not infinite p.p., and satisfies the equation

$$f(x + y) = f(x) + f(y)$$

for all x, $y \in R$. Prove that f is (a) everywhere finite; (b) bounded in every bounded interval; (c) continuous; and (d) of the form $f(x) = ax$, for some real a. Show that (a) to (c) hold also for functions on R^n. What is the result corresponding to (d)?

18. If f is nonzero on the nonmeasurable subset A of E, and zero in $E \setminus A$, show that f cannot be integrable over E.

19. (a) Is it true that, if f^2 and g^2 are integrable, then fg is integrable? (b) Prove that if $(\int fg)^2 = \int f^2 \int g^2$, then f and g are essentially proportional.

20. Prove the Schwarz inequality for complex functions,

$$\left| \int fg \right|^2 \le \int |f|^2 \int |g|^2,$$

(a) using the corresponding result for real functions, and (b) directly.

21. If f is complex and $|\int f| = \int |f|$, what can be said about $\arg f$ ($= \arctan \Im f/\Re f$)? If f is a vector function and $\|\int f\| = \int \|f\|$, what does this imply about f?

22. Prove the inequality $|\int f| \le \int |f|$ for complex functions by considering $f_1 = e^{i\alpha} f$, where α is a suitably chosen real number. Develop an analogous method for vector functions. (A little elementary matrix theory will be needed.)

[4]

The Integral II

4.1. Convergence Theorems

The results of Sec. 3.2 are, generally speaking, to be found in any theory of integration; the advantages of the Lebesgue integral over more elementary definitions (such as Riemann's) were not clearly revealed. In this chapter, on the other hand, most of the theorems are to be found only in the Lebesgue theory. This is particularly true of the convergence theorems of the present section; indeed these are one of the main justifications for introducing the additional complications of Lebesgue's theory as compared with Riemann's.

The type of convergence which is relevant in many situations is (pointwise) *convergence almost everywhere*. "Convergence" should always be understood in this sense, unless the contrary is indicated; $f_r \to f$ will mean $f_r(x) \to f(x)$ for almost all x (in some set E, to be understood from the context). The limit function of a convergent sequence is thus not in general defined everywhere, but it is defined almost everywhere, which is adequate for present purposes.

Although convergence p.p. is obviously less restrictive than uniform convergence, it is in some sense not much less restrictive; more precisely, we have:

THEOREM 4.1a (Egorov). Let E be a set of finite measure, f_r a sequence of measurable functions such that $f_r \to f$ (p.p. in E). Then for any given $\epsilon > 0$ there is a subset F of E such that $m(E \setminus F) < \epsilon$ and $f_r \to f$ uniformly on F.

Proof. It may be assumed that $f_r \to f$ everywhere in E; for, if not, we have only to consider the subset E_0 of E on which $f_r \to f$, in what follows. Write

$$g_r = |f - f_r|, \qquad E_{pq} = \{x: g_r(x) < q^{-1} \text{ for } r \geq p\}.$$

Then, for fixed q, the sets $E_{1q}, E_{2q}, \cdots$ are increasing, and their union is E. By Theorem 2.3h, $m(E_{pq}) \to m(E)$ as $p \to \infty$; there is thus an integer $p(q)$ such that $m(E \setminus E_{p(q)q}) < \epsilon 2^{-q}$. Now let

$$F = \bigcap_{q=1}^{\infty} E_{p(q)q};$$

in F, $g_r < q^{-1}$ if $r \geq p(q)$, that is to say, $g_r \to 0$ uniformly in F. Since $E \setminus F \subset \bigcup E \setminus E_{p(q)q}$, it follows that

$$m(E \setminus F) \leq \sum m(E \setminus E_{p(q)q}) < \epsilon \sum 2^{-q} = \epsilon.$$

The theorem is no longer true if E is of infinite measure. A counter-example is obtained by taking $E = R$, with f_r the characteristic function of $[r, r + 1]$. It is not possible to sharpen the theorem by replacing "$m(E \setminus F) < \epsilon$" by "$m(E \setminus F) = 0$"; counterexamples to this are easy to find.

Extending the definition given in Sec. 3.3, we define a family $(f_i)_{i \in I}$ of functions to be *dominated* in E by g if g is a function, integrable over E, such that $|f_i| \le g$ for all $i \in I$. The type of convergence which appears in the theorem which follows is known as *dominated convergence*.

THEOREM 4.1b (Lebesgue). If f_r is a dominated sequence of integrable functions such that $f_r \to f$ as $r \to \infty$, then f is integrable and $\lim \int |f - f_r| = 0$, so that

$$\int f = \lim \int f_r.$$

Proof. By Lemma 3.3d, f is measurable; if the sequence f_r is dominated by ϕ, then obviously f is dominated by ϕ, and so it is integrable, by Theorem 3.3e. Write $g_r = |f - f_r|$; evidently $g_r \le 2\phi$, and $g_r \to 0$ p.p. If it can be proved that $\int g_r \to 0$, then the theorem will follow, since by Theorem 3.2e we have $|\int f - \int f_r| \le \int g_r$.

Choose any admissible dissection $\{E_1, E_2, \cdots\}$ for ϕ over E, consisting of sets of finite measure. Given $\epsilon > 0$, choose N so that $\sum_{r=N+1}^{\infty} h_r(\phi) \, m(E_r) < \epsilon/4$; write $E' = \bigcup_{r=1}^{N} E_r$. Let $h = \sup' h_r(\phi)$, the supremum being over all r between 1 and N such that $h_r(\phi) < \infty$; h is thus finite. By Theorem 4.1a, $g_r \to 0$ uniformly on a subset E'' of E' such that $m(E' \setminus E'') < \epsilon/8h$. There is therefore an r' such that $g_r < \epsilon/4m(E')$ on E'' if $r \ge r'$. Then, by Lemma 3.2d,

$$\int_E g_r = \int_{E''} g_r + \int_{E' \setminus E''} g_r + \int_{E \setminus E'} g_r.$$

The first integral on the right does not exceed $\epsilon/4$, if $r \ge r'$, because on E'' we have then $g_r < \epsilon/4m(E') < \epsilon/4m(E'')$. The second integral is similarly $< \epsilon/4$, since $g_r \le 2h$ on E' and $m(E' \setminus E'') < \epsilon/8h$. And the third is not greater than $2 \int_{E \setminus E'} \phi < 2 \sum_{r=N+1}^{\infty} h_r(\phi) \, m(E_r) < \epsilon/2$. Hence, if $r \ge r'$,

$$\int_E g_r < \epsilon,$$

which is what we wanted.

COROLLARY. If E is of finite measure, and if f_r is a sequence of integrable functions such that $|f_r| \le a < \infty$ for all r and $f_r \to f$ as $r \to \infty$, then f is integrable and

$$\int_E f = \lim \int_E f_r.$$

The type of convergence which appears in the above corollary is *bounded convergence*; the result itself is *Lebesgue's Theorem on Bounded Convergence*.

In general, some restriction on the sequence f_r is needed for the validity of the relation $\int f = \lim \int f_r$. Consider the functions defined by

$$
\begin{aligned}
f_r(x) &= r^2 x && (0 \le x \le r^{-1}) \\
&= 2r - r^2 x && (r^{-1} < x \le 2r^{-1}) \\
&= 0 && (2r^{-1} < x \le 1);
\end{aligned}
$$

then $f_r \to 0$, but $\int f_r = 1$ for all r, so that $\int \lim f_r \ne \lim \int f_r$ in this case.

The lemma which follows is frequently useful; it shows that integrable functions may be approximated by bounded functions which vanish outside bounded sets. As in Sec. 2.3, $E^{(k)}$ denotes the intersection of the set E with the interval $I^{(k)} = \{x\colon -k \le x_r \le k, 1 \le r \le n\}$. The kth *truncate* of the function f is the function $f^{[k]}$ defined by

$$
\begin{aligned}
f^{[k]}(x) &= f(x) && \text{if } f(x) \le k, \\
&= k && \text{if } f(x) > k.
\end{aligned}
$$

If $f \ge 0$, then $\lim \int_{E^{(p)}} f^{[q]}$ may be finite or infinite, but it is evidently independent of the way in which p, $q \to \infty$.

LEMMA 4.1c. Let f be nonnegative; a necessary and sufficient condition that f should be integrable over E is that $f^{[q]}$ should be integrable over $E^{(p)}$ for all p, q and that $\lim \int_{E^{(p)}} f^{[q]}$ should be finite. If f is integrable, then

$$
\int_E f = \lim \int_{E^{(p)}} f^{[q]}.
$$

Proof. If f is known to be integrable, then $f^{[q]}$ is measurable, and dominated by f, hence integrable over any measurable subset of E. The proof in this direction is completed by using Theorem 4.1b.

Suppose that $\int_{E^{(p)}} f^{[q]}$ $(= a_{pq}$, say) exists for all p, q and that $a = \lim a_{pq}$ is finite. Write

$$
\begin{aligned}
E_r &= \{x\colon 2^r \le f(x) < 2^{r+1}\} && (-\infty < r < \infty), \\
E_{-\infty} &= \{x\colon f(x) = 0\}, && E_\infty = \{x\colon f(x) = \infty\}.
\end{aligned}
$$

Clearly, $m(E_\infty) = 0$; otherwise, we would have $a = \infty$. Also

$$
\sum_{r=q'}^{q} 2^{r+1} m(E_r^{(p)}) \le 2a_{pq} \le 2a;
$$

it follows, making $p \to \infty$ and then $q' \to -\infty$, $q \to \infty$ that

$$
\sum_{r=-\infty}^{\infty} 2^{r+1} m(E_r) \le 2a.
$$

Hence the function which is equal to 0 in $E_{-\infty}$, 2^{r+1} in E_r and ∞ in E_∞ is integrable. Since f is dominated by this function and is measurable (being a limit of measurable functions), it is integrable, by Theorem 3.3e. This completes the proof.

It is, of course, clear that a very concise proof of Lemma 4.1c could be given, using Theorem 3.5c. As corollaries of the lemma, one deduces at once that if $\int_E f$ exists, then it is equal both to lim $\int_E f^{[q]}$ and to lim $\int_{E^{(p)}} f$.

THEOREM 4.1d (Fatou). If $f_r \geq 0$ for all r, and $f = \lim \inf f_r$, then $\int f \leq \lim$ inf $\int f_r$.

Proof. Write $g_r = \inf f_s$ for $s \geq r$; then $f = \lim g_r$. Using Theorem 4.1b, Corollary, we have

$$\int_{E^{(p)}} f^{[q]} = \lim_{r\to\infty} \int_{E^{(p)}} g_r^{[r]}$$

$$= \lim \inf \int_{E^{(p)}} g_r^{[q]} \leq \lim \inf \int_{E^{(p)}} f_r^{[q]} \leq \lim \inf \int_E f_r.$$

Hence

$$\int_E f = \lim_{p,\,q\to\infty} \int_{E^{(p)}} f^{[q]} \leq \lim \inf \int_E f_r.$$

It should, perhaps, be remarked that the convention explained at the beginning of Sec. 3.2 is being maintained; the qualification "if each f_r is integrable" is to be understood in the hypotheses of Fatou's theorem. Similarly, the conclusion, in full, is "if lim inf $\int f_r$ is finite, then f is integrable, and if f is integrable, then $\int f \leq \lim \inf \int f_r$."

The sign in the conclusion of Fatou's theorem may well be strict inequality; an example has already been given, following Theorem 4.1b.

THEOREM 4.1e (B. Levi). If f_r is a monotonic sequence of integrable functions and $f_r \to f$, then

$$\int f = \lim \int f_r.$$

Proof. There is clearly no loss of generality in taking the sequence to be increasing. There is also no loss of generality in assuming that all functions are nonnegative; if not, we have only to consider the sequence $g_r = f_r - f_1$.

If f is integrable, the conclusion follows at once from Theorem 4.1b. If lim $\int f_r$ is finite, then f is integrable, by Theorem 4.1c, and the proof is complete.

As in the case of Lemma 4.1c, a short direct proof can also be given on the basis of Theorem 3.5c and Theorem 2.3h.

It is clear that any of the theorems of this section could be translated immediately into a theorem about infinite series of functions, the sequence of functions appearing in the original version being the sequence of partial sums of the series. Some such translations are included among the exercises at the end of the chapter.

4.2. Fubini's Theorems

The main result (Theorem 4.2b) relates integration over a set in a product space to integration in the factor spaces. A useful by-product is a criterion for the inversion of the order of integration in a repeated integral.

The notation is as follows: E will denote a measurable set in $R^{n+p} = R^n \times R^p$; x is a point in R^n, y in R^p. The section of E by x (Sec. 1.1) will be written $E(x)$. If $f(x, y)$ is defined in R^{n+p}, then $\int f(x, y)\, dy$ (if it exists) is written $g(x)$. The integral here is, of course, over $E(x)$. When $\int g$ is written, the integral is to be taken over any set containing the projection of E on R^n, and similarly for the integral of any function of x. The same symbol m is used for Lebesgue measure in any of the spaces under discussion.

THEOREM 4.2a. *If E is of finite measure in R^{n+p}, then, for almost all x, $E(x)$ is measurable and of finite measure in R^p; $m(E(x))$ is an integrable function of x, and*

$$\int m(E(x))\, dx = m(E).$$

Proof. We prove this in stages, taking sets E which become increasingly complicated.

(1) If E is a bounded interval, the result is obvious. Here $E(x)$ is measurable for all x, and $m(E(x))$ is a constant multiple of the characteristic function of the projection of E on R^n.

(2) If E is a finite union of bounded intervals, the theorem follows at once from (1) by addition.

(3) If E is a (bounded) set in $\mathscr{J}(R^{n+p})$, then, for each x, $E(x)$ is a bounded set in $\mathscr{J}(R^p)$ and is therefore measurable and of finite measure. If $E = \bigcup I_r$, then $E(x) = \bigcup I_r(x)$; writing $E_N = \bigcup_{r=1}^N I_r$, we have $E_N(x) = \bigcup_{r=1}^N I_r(x)$. By Theorem 2.3h, $m(E_N(x)) \to m(E(x))$ and $m(E_N) \to m(E)$ as $N \to \infty$. Since $m(E(x))$ is a limit of measurable functions, it is measurable; it is clearly bounded, hence integrable. The relation $\int m(E(x))\, dx = m(E)$ follows by Theorem 4.1b.

(4) If E is the complement (with respect to a bounded interval) of a set in $\mathscr{J}$, the result follows at once from (3) by subtraction.

(5) If E is a bounded measurable set, then for any $\epsilon > 0$ there are sets $J \supset E$, $K \subset E$ (with $J \in \mathscr{J}$, $\mathscr{C}K \in \mathscr{J}$) such that $m(J) - m(K) < \epsilon$. Take a sequence J_r, K_r of such sets such that $J_r \supset J_{r+1}$, $K_r \subset K_{r+1}$ for all r, and $m(J_r) - m(K_r) \to 0$ as $r \to \infty$. Then, using (3) and (4),

$$\int \{m(J_r(x)) - m(K_r(x))\}\, dx \to 0;$$

since $m(J_r(x)) - m(K_r(x))$ is a decreasing function of r, for each x, the limit as $r \to \infty$ exists, and, by Theorem 3.2h, this limit is 0 p.p. It follows that $E(x)$ is measurable for almost all x, and that $m(E(x)) = \lim m(J_r(x))$ p.p. An application of Theorem 4.1b again completes the proof in this case.

(6) If E is unbounded, (5) applies to each $E^{(k)}$, and if now $k \to \infty$, the required result follows, using Theorem 4.1e.

THEOREM 4.2b. If $f(x, y)$ is integrable over E, then, for almost all x, $f(x, y)$ is integrable over $E(x)$; if $g(x) = \int_{E(x)} f(x, y)\, dy$, then g is integrable and

$$\int g = \int f$$

(or, $\int \{\int f(x, y)\, dy\}\, dx = \iint f(x, y)\, dx\, dy$).

Proof. It is convenient (although not essential) to use the result of Theorem 3.5c, by which the integral of a (nonnegative) function is the measure of its ordinate set. If Ω is the ordinate set of f over E, then $\Omega(x)$ is the ordinate set of $f(x, y)$, as a function of y, over $E(x)$. The required result follows at once from Theorem 4.2a, if f is nonnegative. If f is general, it is only necessary to write $f = f^+ - f^-$, and apply the result just proved to f^+ and f^- separately.

It is clear that the parts played by x and y in the above two theorems could be interchanged.

The next problem is to obtain conditions under which the order of integration in a repeated integral may be changed. Consider the following two examples; in each case x and y are real variables and the integrals with respect to each variable are over $[0, 1]$:

(1) $f(x, y) = (x^2 - y^2)/(x^2 + y^2)^2$; here

$$\int \{\int f(x, y)\, dy\}\, dx = \int (1 + x^2)^{-1}\, dx = \pi/4,$$

$$\int \{\int f(x, y)\, dx\}\, dy = -\int (1 + y^2)^{-1}\, dy = -\pi/4.$$

(We are assuming that the Lebesgue integral obeys the usual rules of elementary calculus; we have not proved anything in this direction as yet.)

(2) Let $\mathscr{F}$ be a class of subsets of $[0, 1]$ of measure zero, such that (a) given F, $F' \in \mathscr{F}$ either $F \subset F'$ or $F \supset F'$; (b) the union A of all the sets in $\mathscr{F}$ is not of measure zero. (The existence of such classes is a consequence of Zorn's maximal principle, which we do not propose to discuss.) Write, for x, $y \in A$, $x\rho y$ if there is a set F in $\mathscr{F}$ such that $x \in F$, $y \notin F$. Now let

$$B = \{(x, y): x\rho y\}$$

and let $f(x, y)$ be the characteristic function of B. Then for each y the section of B by y is of measure zero, and for each $x \in A$ the section of B by x is $A \setminus Z$

for some set Z of measure zero. It follows that $\int\{\ \int f(x, y)\,dx\}\,dy$ exists and is zero, while $\int\{\{\int f(x),y)\,dy\}\,dx$ either exists and is equal to $(m(A))^2$ (if A is measurable) or fails to exist.

It is clear from the above that some condition on f, other than the mere existence of one repeated integral (or even both), is needed to make the change of order of integration valid. Perhaps the most useful condition is contained in the following theorem:

THEOREM 4.2c. *If $f(x, y)$ is measurable (in R^{n+p}) and if $\int\{\int |f(x, y)|\,dx\}\,dy$ exists, then*

$$\int\{\int f(x, y)\,dx\}\,dy = \int\{\int f(x, y)\,dy\}\,dx.$$

Proof. Since f is measurable, so is $|f|$, and hence so is $|f|^{[r]}$ for all r. The integral of $|f|^{[r]}$ over $E^{(q)}$ exists for all q, r and, by Theorem 4.2b, is equal to the corresponding repeated integral (first with respect to x, then y), which clearly does not exceed $\int\{\int |f(x, y)|\,dx\}\,dy$. Applying Lemma 4.1c, $\int |f|$ exists; by Theorem 3.3e, $\int f$ exists, and the result now follows from Theorem 4.2b.

The last result of this section, although somewhat special, is quite important. All integrals are supposed to be over R^n; by the invariance of Lebesgue measure under translations, if one of $\int f(x - y)\,g(y)\,dy$, $\int f(y)\,g(x - y)\,dy$ exists, so does the other, and the two are equal. The resulting function of x is the *convolution product* of f and g, and will be denoted by $f \circ g$ or $g \circ f$.

THEOREM 4.2d. *If f, g are integrable, then $f \circ g$ exists and is integrable; and*

$$\int |f \circ g| \le \int |f| \int |g|.$$

Proof. It is evident that g is measurable as a function of (x, y); by Theorem 2.3j, $f(x - y)$ is measurable, and so $f(x - y)\,g(y)$ is measurable. The integral $\int\{\int |f(x - y)\,g(y)|\,dx\}\,dy$ evidently exists and is equal to $\int |f| \int |g|$; hence, by Theorem 4.2c, $\int\{\int |f(x - y)\,g(y)|\,dy\}\,dx$ also exists and has the same value. The fact that $\int |\int f(x - y)\,g(y)\,dy|\,dx\,(= \int |f \circ g|)$ exists and does not exceed $\int |f| \int |g|$ is then immediate.

4.3. Approximations to Integrable Functions

In this section all integrals will be supposed to be over R^n; as explained in Sec. 3.1, this involves no loss of generality. We shall say that an integrable function f is *approximable* (can be approximated) by functions of some class $\mathscr{F}$ if, for any given $\epsilon > 0$, there is a function $\phi \in \mathscr{F}$ such that $\int |f - \phi| < \epsilon$. We have already had one result of this type; Lemma 4.1c shows that any integrable function is approximable by bounded functions of compact support (that is, zero outside some bounded set). It will appear that integrable functions can be approximated by functions whose behavior is very regular indeed (Theorem 4.3b). In the first theorem, "step function" is to be taken in the strict sense, as a finite linear combination of characteristic functions of bounded intervals.

THEOREM 4.3a. *Any integrable function is approximable by step functions.*

Proof. By considering a suitable dissection of R^n, consisting of bounded sets, it is clear (see Lemma 4.4d) that any integrable function can be approximated by finite linear combinations of characteristic functions of bounded measurable sets. The required result will follow if it can be shown that the characteristic function of a bounded measurable set is approximable by finite linear combinations of characteristic functions of intervals.

Given E, there exists $J \supset E$ with $m(J) - m(E) < \epsilon/2$; and a finite union of intervals $J' \subset J$ with $m(J) - m(J') < \epsilon/2$. Then evidently $\int |\chi_E - \chi_{J'}| \leq \int |\chi_E - \chi_J| + \int |\chi_J - \chi_{J'}| < \epsilon/2 + \epsilon/2 = \epsilon$, which is what was wanted.

A function is said to be *infinitely differentiable* if its derivatives of all orders exist. Each derivative must be continuous, and, in particular, the function itself is continuous.

THEOREM 4.3b. *Any integrable function is approximable by infinitely differentiable functions of compact support.*

Proof. In view of Theorem 4.3a, it is enough to prove that the characteristic function of a bounded interval I can be approximated in this way. Suppose first that $n = 1$; let I be (a, b) and let $\epsilon > 0$ be given. Define ϕ by

$$
\begin{aligned}
\phi(x) &= 0 && \text{if } x \leq a - \epsilon/2 \\
&= \exp\left(4\epsilon^{-2} - 4(\epsilon^2 - 4(x-a)^2)^{-1}\right) && \text{if } a - \epsilon/2 < x \leq a \\
&= 1 && \text{if } a < x < b \\
&= \exp\left(4\epsilon^{-2} - 4(\epsilon^2 - 4(b-x)^2)^{-1}\right) && \text{if } b \leq x < b + \epsilon/2 \\
&= 0 && \text{if } b + \epsilon/2 \leq x.
\end{aligned}
$$

Then it is easy to verify that ϕ is infinitely differentiable, and that $\int |\chi_I - \phi| < \epsilon$.

If $n > 1$, let I be the product of intervals $I_r = (a_r, b_r)$ in R. Choose ϵ' so that $(1 + \epsilon')^n \leq 1 + \epsilon$. For each r, construct a function ϕ_r as above, with a_r, b_r, ϵ' in place of a, b, ϵ. Then the function $\phi(x) = \phi_1(x_1) \phi_2(x_2) \cdots \phi_n(x_n)$ has clearly all the required properties.

The next theorem makes use of a weakened form of the result just proved:

THEOREM 4.3c. If f is integrable, then $\int |f(x+y) - f(x)| \, dx \to 0$ as $y \to 0$.

Proof. Given $\epsilon > 0$, there is a continuous function ϕ of compact support such that $\int |f(x) - \phi(x)| \, dx < \epsilon/3$ and $\int |f(x+y) - \phi(x+y)| \, dx < \epsilon/3$. Suppose that the support of ϕ is contained in a sphere of radius ρ with center at the origin; let S be the sphere with the same center and radius $\rho + 1$. Since, by Theorem 1.5c, ϕ is uniformly continuous, there is a real number δ (which may be supposed < 1) such that $|\phi(x+y) - \phi(x)| < \epsilon/3m(S)$ whenever $d(y, 0) < \delta$. Then

$$\int |\phi(x) - \phi(x+y)| \, dx = \int_S |\phi(x) - \phi(x+y)| \, dx < m(S). \ \epsilon/3m(S) = \epsilon/3;$$

and so $\int |f(x+y) - f(x)| \, dx < \epsilon$ whenever $d(y, 0) < \delta$, which is the required result.

THEOREM 4.3d. If the integrable function f and the positive real number ϵ are given, there is a positive real δ such that

$$\int |f - f \circ g| < \epsilon$$

for any nonnegative function g whose support is in the sphere with center at the origin and of radius δ, and which satisfies the condition $\int g = 1$.

Proof. Let ϕ, S, and δ be as in the previous theorem. Let S' be the sphere with center at the origin and radius δ. Then, for any x,

$$|\phi(x) - \phi \circ g(x)| = |\int (\phi(x) - \phi(y)) g(x - y) \, dy| < \epsilon/3m(S),$$

whenever g satisfies the conditions of the theorem, since the integrand is nonzero only if $x - y \in S'$, and, in that case, it is less than $\epsilon g(x - y)/3m(S)$ in absolute value. Then

$$\int |f - f \circ g| \le \int |f - \phi| + \int |\phi - \phi \circ g| + \int g \circ |f - \phi|;$$

the first term on the right is less than $\epsilon/3$, by the choice of ϕ, and the third is less than $\epsilon/3$ by Theorem 4.2d. The integrand in the second term does not exceed $\epsilon/3m(S)$ in absolute value, and its support is evidently contained in S. Hence the second term also is less than $\epsilon/3$, and the theorem is proved.

THEOREM 4.3e. Any integrable function is approximable by linear combinations of convolution products of the form $f \circ f$. Any complex function is approximable by linear combinations of products of the form $f \circ f$. The functions which occur in the products may be assumed to be infinitely differentiable and of compact support.

Proof. This follows from the previous theorem, in view of the identities

$$f \circ g = \{(f+g) \circ (f+g) - f \circ f - g \circ g\}/2,$$
$$= \{(f+g) \circ (f+g) - (f-g) \circ (f-g) + i(f+ig) \circ (f-ig)$$
$$- i(f-ig) \circ (f+ig)\}/4,$$

for the general and complex cases, respectively. (The function g in the complex analogue of Theorem 4.3d is still real.) The last assertion of the theorem is a consequence of Theorem 4.3b and the result just proved.

In the course of the proof just given, the fact that the convolution product is commutative ($f \circ g = g \circ f$) is essential; this is, of course, simply a reformulation of the property mentioned in Sec. 4.2, when the definition was first given.

4.4. The L_p Spaces

Let p be a positive real number. The measurable functions f such that $\int_E |f|^p$ exists form the class of pth power integrable functions over E, denoted by $L_p(E)$, or simply L_p if no confusion is likely. If $p \geq 1$ the classes L_p have more agreeable properties than if $0 < p < 1$, and we assume $p \geq 1$ from now on. Indeed, the cases $p = 1$, $p = 2$ are the only ones which we consider in detail; $p = 1$ gives, of course, the class of Lebesgue-integrable functions.

It is obvious that if $f \in L_p$, and c is a constant, then $cf \in L_p$. Also, since $|f+g|^p \leq 2^p(|f|^p + |g|^p)$, if $f, g \in L_p$ then $f+g \in L_p$. In other words, L_p is a linear space. For any $p \geq 1$ and any $f, g \in L_p$, we have

$$(\int |f+g|^p)^{1/p} \leq (\int |f|^p)^{1/p} + (\int |g|^p)^{1/p}$$

(*Minkowski's inequality*). For $p = 1$ this is trivial, and for $p = 2$ it is an immediate consequence of Schwarz's inequality (Lemma 3.4a), as can be seen on squaring both sides. For general p it is a consequence of an inequality due to Hölder which generalizes that of Schwarz.

If $\|f\|_p$ is written for $(\int |f|^p)^{1/p}$, it is clear, from the definition and from Minkowski's inequality, that (1) $\|cf\|_p = |c| \|f\|_p$, (2) $\|f+g\|_p \leq \|f\|_p + \|g\|_p$. That is, $\|\cdot\|_p$ is a seminorm on L_p. It is not a norm, since $\|f\|_p = 0$ implies only that $f = 0$ p.p., not that $f = 0$ in the strict sense. If one considers the space $\mathscr{L}_p$ of equivalence classes of functions in L_p modulo functions which are almost everywhere zero, then the seminorm on L_p induces a genuine norm on $\mathscr{L}_p$. With distance d defined in L_p by $d(f, g) = \|f-g\|_p$, L_p is a semimetric space, while $\mathscr{L}_p$, with the corresponding distance, is a metric space in the strict sense. As is customary, however, the distinction between L_p and $\mathscr{L}_p$ will not be insisted on.

Since $||f||_p \to$ ess sup $|f|$ as $p \to \infty$, if the limit exists at all, it is natural to define L_∞ to be the class of essentially bounded measurable functions, with $||f||_\infty =$ ess sup $|f|$. This is again a seminormed linear space. It enjoys some —though not all—of the properties of the spaces L_p with $1 \leq p < \infty$. For certain purposes it is preferable to consider the subspace of L_∞ consisting of functions which essentially vanish at infinity; we do not go into details.

Many of the results already obtained generalize from L_1 to L_p. For instance, it is natural to define a function f (or a family of functions f_i) to be p *dominated* if there is a function $g \in L_p$ such that $|f| \leq g$ (or $|f_i| \leq g$ for all i). Then it follows at once, from Theorem 3.3e, that a function f is in L_p if and only if it is measurable and p dominated. Also:

THEOREM 4.4a. If f_r is a p-dominated sequence of functions in L_p which tends to f p.p., then $f \in L_p$ and

$$||f - f_r||_p \to 0 \text{ as } r \to \infty.$$

Proof. If f_r is p dominated by ϕ, then $|f - f_r|$ is p dominated by 2ϕ, and so $|f - f_r|^p$ is a dominated sequence of integrable functions which $\to 0$ p.p. The result now follows at once from Theorem 4.1b.

THEOREM 4.4b. If $f \in L_1$, $g \in L_p$, then $f \circ g \in L_p$, and

$$||f \circ g||_p \leq ||f||_1 \, ||g||_p.$$

Proof. We have already had the case $p = 1$ (Theorem 4.2d). The proof for $p = 2$ is as follows. By Theorem 4.2d, $\int f(y) g^2(x - y) \, dy$ exists (for almost all x); since f is integrable, it follows, from Lemma 3.4a, that $(f \circ g)(x) = \int f(y) g(x - y) \, dy$ exists for almost all x; and moreover,

$$|(f \circ g)(x)|^2 \leq \int |f| \int |f(y) g^2(x - y)| \, dy.$$

The function on the right is an integrable function of x, by Theorem 4.2d, hence $f \circ g \in L_2$. Integrating with respect to x, and applying Theorem 4.2d to the convolution product $f \circ g^2$, we have $||f \circ g||_2^2 \leq ||f||_1^2 \, ||g||_2^2$, which completes the proof.

The results of Sec. 4.3 on approximations can also be generalized. A function $f \in L_p$ is said to be p *approximable* by functions $\phi \in \mathscr{F}$ ($\subset L_p$) if for any $\epsilon > 0$ there is a $\phi \in \mathscr{F}$ such that $||f - \phi||_p < \epsilon$ (or, what is the same thing, such that $\int |f - \phi|^p < \epsilon$). In the language of topology, if every $f \in L_p$ can be approximated by functions of $\mathscr{F}$, then $\mathscr{F}$ is dense in L_p. The theorems of Sec. 4.3 could be stated in terms of the density of certain classes of functions in L_1.

The principles already used in Sec. 4.3: (1) if f is approximable by functions $\phi \in \mathscr{F}$, and every $\phi \in \mathscr{F}$ is approximable by functions $\psi \in \mathscr{G}$, then f is approximable by functions $\psi \in \mathscr{G}$; (2) if f is approximable by (finite) linear combinations of functions in $\mathscr{F}$, and $\mathscr{G}$ is as in (1), then f is approximable by (finite) linear combinations of functions in $\mathscr{G}$, remain valid here for p approximation, in view of Minkowski's inequality. A more special result is:

LEMMA 4.4c. If $f \in L_1 \cap L_p$ is bounded, and can be approximated by functions in the uniformly bounded class $\mathscr{F} \subset L_1 \cap L_p$, then it can be p approximated.

Proof. Suppose that K is such that $|\phi| \leq K$ for all $\phi \in \mathscr{F}$, and $|f| \leq K$. Then $|f - \phi| \leq 2K$, and

$$|f - \phi|^p \leq |f - \phi| (2K)^{p-1},$$

so that, on integrating and taking the pth root,

$$\|f - \phi\|_p \leq \|f - \phi\|_1^{1/p} (2K)^{1-1/p},$$

which yields the desired conclusion.

LEMMA 4.4d. Any $f \in L_p$ is p approximable by finite linear combinations of characteristic functions of bounded measurable sets.

Proof. It may be assumed that $f \geq 0$, without loss of generality. Consider a dissection $\mathscr{E}$ consisting of bounded sets, such that

$$\sum B_r(f^p) \, m(E_r) - \int f^p < \epsilon/2,$$

and let N be such that

$$\sum_{r=N+1}^{\infty} B_r(f^p) \, m(E_r) < \epsilon/2.$$

Write χ_r for the characteristic function of E_r; let

$$\phi = \sum_{r=1}^{\infty} B_r(f) \chi_r, \qquad \phi_N = \sum_{r=1}^{N} B_r(f) \chi_r.$$

Using the elementary inequality $|a - b|^p \leq |a^p - b^p|$, valid for all $a \geq 0$, $b \geq 0$ and $p \geq 1$, we have

$$\int |f - \phi_N|^p \leq \int |f^p - \phi_N^p|$$
$$\leq \int |f^p - \phi^p| + \int |\phi^p - \phi_N^p|$$
$$< \epsilon/2 + \epsilon/2 = \epsilon.$$

This completes the proof.

THEOREM 4.4e. Any $f \in L_p$ is p approximable by (1) step functions, (2) infinitely differentiable functions of compact support, (3) convolution products of the form $f \circ g$, and (4) linear combinations of convolution products of the form $h \circ h$ (or, in the complex case, $h \circ \bar{h}$). The functions which occur in the convolution products in (4) may be assumed to be infinitely differentiable and of compact support.

Proof. These results follow at once from the corresponding results in Sec. 4.3, since, in each case, the conditions of Lemma 4.4c are satisfied.

4.5. Convergence in Mean

There is another type of convergence, besides convergence almost everywhere, which arises naturally in integration theory. We have had instances of it already, in Theorem 4.1b and its extension, Theorem 4.4a. Let $f, f_1, f_2, \cdots \in L_p$; the sequence f_r *converges* to f *in mean with index* p (is p convergent) if $\| f - f_r \|_p \to 0$ as $r \to \infty$. The sequence f_r is a *Cauchy sequence* in mean with index p (is a p Cauchy sequence) if $\| f_r - f_s \|_p \to 0$ as $r, s \to \infty$.

By Theorem 4.4a, p-dominated convergence almost everywhere implies p convergence. In the absence of domination the implication fails, as is shown by the example following Theorem 4.1b. There is no implication in the other direction; indeed, a p-convergent sequence may be divergent everywhere. Consider the sequence f_r, where f_r is the characteristic function of the interval $[r2^{-k} - 1, (r + 1) 2^{-k} - 1]$ for $2^k \leq r < 2^{k+1}$. It is clear that $\int f_r \to 0$ as $r \to \infty$, but $f_r(x)$ does not tend to a limit for any x in $[0, 1]$; for any given x, r' there exist $r, s > r'$ with $f_r(x) = 0, f_s(x) = 1$.

It will be noted that, in the example just given, it is possible to select a subsequence of the given sequence which is convergent p.p. It appears in the course of the proof of the next theorem that this is indeed the general situation also.

The theorem which follows contains the essential part of the Riesz-Fischer theorem on Fourier series (Theorem 4.6c) and is sometimes given the same name. In topological terms it asserts that L_p (and also $\mathscr{L}_p$) is complete as a semimetric (or metric) space.

THEOREM 4.5a. If f_r is a p Cauchy sequence in L_p, there is an essentially unique function f in L_p such that f_r converges to f in mean with index p.

Proof. Given $\epsilon > 0$ there is an integer $N(\epsilon)$ such that

$$\int | f_r - f_s |^p < \epsilon^{p+1} \quad \text{if} \quad r, s \geq N(\epsilon).$$

Write $E(\epsilon; r, s) = \{x: | f_r - f_s | \geq \epsilon\}$; then $m(E(\epsilon; r, s)) < \epsilon$ if $r, s \geq N(\epsilon)$. Writing N_k for $N(2^{-k}\epsilon)$, it may be assumed, without loss of generality, that

$N_1 < N_2 < N_3 < \cdots$. If $E_k = E(\epsilon 2^{-k}; N_{k+1}, N_k)$, then $m(E_k) < 2^{-k}\epsilon$ for all k; and if $F_K = \bigcup_{k=K}^{\infty} E_k$, then $m(F_K) \leq \epsilon \sum_{k=K}^{\infty} 2^{-k} = \epsilon 2^{1-K}$.

Since, for fixed k, $|f_{N_{k+1}}(x) - f_{N_k}(x)| < \epsilon 2^{-k}$ for all $x \notin E_k$, this inequality holds for all $k \geq K$ if $x \notin F_K$. Now, $\sum_{k=1}^{\infty} |f_{N_{k+1}}(x) - f_{N_k}(x)|$ fails to converge only if, for some $\delta > 0$ and arbitrarily large k, $|f_{N_{k+1}}(x) - f_{N_k}(x)| \geq \delta$, which implies $x \in F_K$ for all K.

Thus, $x \in G = \bigcap_{K=1}^{\infty} F_K$, and $m(G) \leq m(F_K) = \epsilon 2^{1-K}$ for all K, so that $m(G) = 0$. It follows that $\sum_{k=1}^{\infty} (f_{N_{k+1}} - f_{N_k})$ is convergent p.p. Let the sum of the series be f; evidently, $f_{N_k} \to f$ as $k \to \infty$. Also, $|f_{N_k} - f_r|^p \to |f - f_r|^p$ p.p. as $k \to \infty$.

By Fatou's theorem (Theorem 4.1d), since $\liminf \int |f_{N_k} - f_r|^p < \epsilon$ if $r \geq N(\epsilon)$, $|f - f_r|^p$ is integrable and $\int |f - f_r|^p < \epsilon$ if $r \geq N(\epsilon)$. That is, f_r is p convergent to f. Since $f_r \in L_p$, $f - f_r \in L_p$, it follows that $f \in L_p$.

The essential uniqueness of f is easy to prove. Suppose that f_r is p convergent to f and to g; by Minkowski's inequality,

$$\left(\int |f - g|^p\right)^{1/p} \leq \left(\int |f - f_r|^p\right)^{1/p} + \left(\int |f_r - g|^p\right)^{1/p};$$

the right side of this tends to zero as $r \to \infty$, and so the left side is zero, which implies that $f = g$ p.p.

4.6. Fourier Theory

One of the earliest applications of the Lebesgue integral in analysis was to the theory of Fourier series and integrals; the advantages over the Riemann integral are very marked in some directions. We give here only a few specimen results.

It is convenient to use the complex form of Fourier series and integrals, where exponentials appear instead of sines and cosines. All results could, of course, be presented in the real form. Let $f \in L_1(R^n)$; the *Fourier transform* $\mathfrak{F}f$ of f is the function on R^n defined by

$$\mathfrak{F}f(y) = \int e^{ixy} f(x)\, dx.$$

In this, if $n > 1$, xy is to be taken as $x_1 y_1 + \cdots + x_n y_n$; the normalizing factor $(2\pi)^{-n/2}$, which is usually included in the definition, has been omitted for simplicity. Formally, $\mathfrak{F}$ is a *linear operation* on L_1; $\mathfrak{F}(cf) = c\mathfrak{F}f$, and $\mathfrak{F}(f + g) = \mathfrak{F}f + \mathfrak{F}g$.

THEOREM 4.6a. If $f \in L_1$, $\mathfrak{F}f(y)$ is a continuous function of y; $||\mathfrak{F}f||_\infty \leq ||f||_1$, and $\mathfrak{F}f(y) \to 0$ as $||y|| \to \infty$.

Proof. Evidently $\mathfrak{F}f(y)$ exists for each $y \in R^n$, by Theorem 3.3e. Continuity may be proved as follows. Given $\epsilon > 0$, there is a function ϕ of compact support

K such that $\int_K |f - \phi| < \epsilon/4$, by Lemma 4.1c. Since the function e^{ixy} is continuous for each $x \in K$, it is uniformly continuous, by Theorem 1.5c. It follows that there is an open sphere S with center y such that

$$| e^{ixy} - e^{ixz} | < \epsilon/2\|\phi\|_1$$

for $z \in S$ and all $x \in K$. Then, if $z \in S$,

$$| \mathfrak{F}(f(y) - \mathfrak{F}f(z) | \leq \int | e^{ixy} - e^{ixz} | |f(x)| \, dx$$

$$\leq \int | e^{ixy} - e^{ixz} | |f(x) - \phi(x)| \, dx + \int | e^{ixy} - e^{ixz} | |\phi(x)| \, dx$$

$$\leq 2\|f - \phi\|_1 + \|\phi\|_1 \epsilon/2\|\phi\|_1$$

$$\leq 2\epsilon/4 + \epsilon/2$$

$$= \epsilon,$$

which establishes the continuity.

Since, by Theorem 3.2e, for each y

$$| \mathfrak{F}f(y) | \leq \int |f| = \|f\|_1,$$

and $\mathfrak{F}f$, being continuous, is measurable, $\mathfrak{F}f \in L_\infty$ and

$$\| \mathfrak{F}f \|_\infty \leq \|f\|_1.$$

Suppose now that f is the characteristic function of an interval $I = \mathsf{X}_{r=1}^n I_r$, where $I_r = (a_r, b_r)$. Then evidently (using a little elementary calculus, justified in Chap. 5),

$$\mathfrak{F}f(y) = \prod_{r=1}^n (iy_r)^{-1} (\exp ib_r y_r - \exp ia_r y_r),$$

and this certainly tends to zero as $\|y\| \to \infty$; each factor is bounded, and, if $\|y\| \to \infty$, then $|y_r| \to \infty$ for at least one r. It follows that $\mathfrak{F}f(y) \to 0$ as $y \to \infty$ if f is a step function (a finite linear combination of characteristic functions). Finally, let f be a general integrable function; given $\epsilon > 0$ there is a step function ϕ such that $\int |f - \phi| < \epsilon/2$ (Lemma 4.1c), whence $| \mathfrak{F}f(y) - \mathfrak{F}\phi(y) | < \epsilon/2$ for all y. And, by what has just been proved, there is a real number a such that $| \mathfrak{F}\phi(y) | < \epsilon/2$ if $\|y\| > a$; therefore,

$$| \mathfrak{F}f(y) | < \epsilon \quad \text{if } \|y\| > a,$$

which shows that $\mathfrak{F}f(y) \to 0$ as $\|y\| \to \infty$.

The last part of the above theorem is usually called the *Riemann-Lebesgue lemma*.

The next result shows that the Fourier transform is a linear operation of a rather special kind, which converts convolution products into ordinary products:

THEOREM 4.6b. If $f, g \in L_1$, then $\mathfrak{F}(f \circ g)(y) = \mathfrak{F}f(y)\, \mathfrak{F}g(y)$.

Proof. $\mathfrak{F}(f \circ g)(y) = \int (e^{ixy} \int f(x - z)\, g(z)\, dz)\, dx$

$$= \int (\int e^{izy}\, g(z)\, e^{i(x-z)y}\, f(x - z)\, dz)\, dx$$

$$= \int (e^{izy}\, g(z) \int e^{i(x-z)y}\, f(x - z)\, dx)\, dz$$

by Theorem 4.2c (the conditions are easily seen to be satisfied)

$$= \int e^{izy}\, g(z)\, \mathfrak{F}f(y)\, dz$$

$$= \mathfrak{F}f(y)\, \mathfrak{F}g(y).$$

For the last theorem of the section, it is convenient to take $n = 1$ and to consider Fourier series rather than Fourier integrals. The restriction to one dimension is not serious, being mainly a matter of notational convenience, but the case of Fourier series is essentially simpler than that of Fourier integrals. Let f be integrable over $(0, 2\pi)$; the *Fourier coefficients* c_k of f are defined, for all integers k, by

$$c_k = (2\pi)^{-1} \int e^{-ikx}\, f(x)\, dx.$$

(We have retained the traditional factor and sign here.) It follows at once, from Theorem 4.6a, that $c_k \to 0$ as $k \to \pm \infty$.

THEOREM 4.6c. If $f \in L_2(0, 2\pi)$ and its Fourier coefficients are c_k, then

$$\sum_{k=-\infty}^{\infty} |c_k|^2 \le \int |f|^2;$$

and if c_k is any sequence of complex numbers such that $\sum |c_k|^2 < \infty$, there is a function in L_2 which has the c_k as its Fourier coefficients.

Proof. The function f, being in L_2, is in L_1 also, since the measure of $(0, 2\pi)$ is finite; the Fourier coefficients are therefore well defined. The first assertion of the theorem follows at once from the identity

$$\int |f - \sum_{k=M}^{N} c_k e^{ikx}|^2\, dx = \int |f|^2 - \sum_{k=M}^{N} |c_k|^2,$$

and the observation that the left side of this is nonnegative.

Given a sequence c_k as in the statement of the theorem, let $f_N = \sum_{k=-N}^{N} c_k e^{ikx}$; then it is clear that $\| f_M - f_N \|_2^2 = \sum_{k=-M}^{-N-1} |c_k|^2 + \sum_{k=N+1}^{M} |c_k|^2 \to 0$ as M,

$N \to \infty$. By Theorem 4.5a, there is a function $f \in L_2$ such that $|| f - f_N ||_2 \to 0$ as $N \to \infty$. Denoting the Fourier coefficients of f by c'_k, we have, for $N > |k|$,

$$| c_k - c'_k | \leq \int | e^{ikx} ||f(x) - f_N(x) | \, dx$$

$$\leq (2\pi)^{1/2} ||f - f_N ||_2$$

(by Lemma 3.4a), and this tends to zero as $N \to \infty$, so that $c'_k = c_k$ for all k.

There is a more refined version of Theorem 4.6c, which asserts that $\Sigma | c_k |^2 = \int |f|^2$, and that the function with given c_k as Fourier coefficients is essentially unique. The proof is not difficult, but it involves considerations which have no very close connection with Lebesgue integration.

There is an integral analogue of Theorem 4.6c, to the effect that if $f \in L_2(-\infty, \infty)$ then $\mathfrak{F}f \in L_2(-\infty, \infty)$, and moreover $|| f ||_2 = (2\pi)^{-1} || \mathfrak{F}f ||_2$. This involves some preliminary study of what is meant by $\mathfrak{F}f$, however, if $f \in L_2$ (it is clear that the defining integral may fail to exist as a Lebesgue integral, for all y), and we do not attempt a treatment here.

EXERCISES

1. Given $\epsilon > 0$, prove that $x^r \to 0$ as $r \to \infty$ uniformly on a subset of $(0, 1)$ of measure greater than $1 - \epsilon$, but not on any subset of $(0, 1)$ of measure 1.

2. Prove that if $m(E) = \infty$, and $f_r \to f$ p.p. in E, then for any $P > 0$ there is a subset F of E with $m(F) > P$ on which $f_r \to f$ uniformly. Show that, in general, there is *not* a subset F of infinite measure on which $f_r \to f$ uniformly.

3. Show that Theorem 4.1a is no longer true if the condition that the functions f_r be measurable is dropped, even if $m(E \setminus F)$ is replaced by $m_*(E \setminus F)$. (Consider functions suitably related to the nonmeasurable sets A_r described in Sec. 2.5.)

4. Produce a sequence of functions f_r, defined on $(0, 1)$, such that $f_r \to 0$ p.p. and $\int f_r \to 0$, but such that the sequence is not dominated in $(0, 1)$.

5. If $f_r \to f$ dominatedly, and g is bounded and measurable, prove that $\int f_r g \to \int fg$. If $g_r \to g$ boundedly, show that $\int f_r g_r \to \int fg$. Is the last result true if $g_r \to g$, not necessarily boundedly, and $\lim \int f_r g_r$ exists?

6. If $f_r \to f$ boundedly, and g is integrable, prove that $\int f_r g \to \int fg$.

7. We define, for real or complex f,

$$f^{\{k\}} = f \quad \text{if} \quad |f| \leq k;$$
$$= k \arg f \quad \text{if} \quad |f| > k.$$

Prove that if $\int_E f$ exists then it is equal to $\lim \int_E f^{\{k\}}$. If $\lim \int_E f^{\{k\}}$ exists, does $\int_E f$ necessarily exist?

8. Show that Theorem 4.1d is no longer true if the condition $f_r \geq 0$ is dropped.

9. If $u_r \geq 0$ p.p., for all r, then

$$\int \sum u_r = \sum \int u_r,$$

in the sense that if either side is finite so is the other, and the two are equal.

10. Prove that if $\sum \int_E |u_r| < \infty$ then $u_r \to 0$ p.p. in E.

11. Show that if $|u_r(x)| \leq r^{-2}$ for all r and all $x \in E$, and v is integrable over E, then $\int \sum u_r v = \sum \int u_r v$.

12. If $u_r(x) = r^{-2} x^{-1/2} \cos rx^{-1}$, show that $\int \sum u_r = \sum \int u_r$, where the integrals are over $(0, 1)$.

13. If $u_r(x) = r^{-1} x^{-1} \sin rx^{-1}$, is it true that $\int \sum u_r = \sum \int u_r$ (a) over $(1/2, 1)$? (b) over $(0, 1)$?

14. If $u_r(x) = e^{-rx} - 2e^{-2rx}$, show that $\int \sum u_r \neq \sum \int u_r$, where the integrals are over $(0, \infty)$.

15. Calculate the convolution product $f \circ g$ and verify directly Theorem 4.2d if $f(x) = e^{-a|x|}$, $g(x) = e^{-b|x|}$ $(a > 0, b > 0)$.

16. Prove that convolution products in L_1 are associative: $f \circ (g \circ h) = (f \circ g) \circ h$.

17. Show that if $f, g \in L_2$, then $f \circ g \in L_\infty$ and $\| f \circ g \|_\infty \leq \| f \|_2 \| g \|_2$.

18. Show that if $f \in L_1, g \in L_\infty$, then $f \circ g \in L_\infty$, and $\| f \circ g \|_\infty \leq \| f \|_1 \| g \|_\infty$. Show also that if g is continuous, then $f \circ g$ is continuous.

19. Produce examples of functions $f, g \in L_1$ such that (a) $f \circ g$ is not continuous (consider suitable modifications of $x^{-1/2}$); (b) $f \circ g$ is not bounded.

20. If $1 \leq p < q < \infty$, and $r = (p + q)/2$, prove that

$$\| f \|_r^{2r} \leq \| f \|_p^p \| f \|_q^q.$$

Deduce that if $f \in L_p$, $f \in L_q$ and $p < r < q$, then $f \in L_r$ also.

21. Prove that if $f \in L_2(R^n)$, then $\int |f(x + h) - f(x)|^2 \, dx \to 0$ as $h \to 0$.

22. Prove that if $f \in L_1$ is not zero p.p., then there is a continuous function g of compact support such that $\int fg \neq 0$.

23. Verify Theorem 4.6b for the functions of Exercise 15.

24. Show that if $f \in L_1(0, 2\pi)$ has Fourier coefficients c_k it is not, in general, true that $\sum |c_k| < \infty$. If $\sum |c_k| < \infty$, is there necessarily a function in L_1 which has the c_k as its Fourier coefficients?

[5]

Calculus

5.1. Change of Variables

In this section a proof is given of the formula for change of variables in a multiple integral. It should, perhaps, be remarked that the essential difficulties of the proof are not really connected with the Lebesgue integral as such, but would appear in any theory of integration. The conditions under which the result is proved are more restrictive than is really necessary; even so, the proof is complicated. The case $n = 1$ is naturally simpler in that considerations involving linear transformations become superfluous.

Let g be a function defined on some subset G of R^n, with values also in R^n: $g(x) = (g_1(x), \cdots, g_n(x))$. It is assumed that G is open and that g and its first partial derivatives are continuous. It is convenient to write $g_{ij}(x)$ for $\partial g_i(x)/\partial x_j$. It follows, from Theorem 1.5c, that g_i and g_{ij} are uniformly continuous functions in any compact subset of G. It is assumed also that g is a 1-1 map of G on to $g(G)$, and that the inverse function g^{-1} is continuous and has continuous first partial derivatives. Under these circumstances g and its first partial derivatives are uniformly continuous in any bounded subset of G. Moreover, if the *Jacobian* $J(x)$ of the function g is defined by

$$J(x) = |\det g_{ij}(x)|$$

(that is, the absolute value of the determinant of the first partial derivatives), then $J(x) \neq 0$ in G. The above assumptions are not independent, but it seems desirable to write them all out explicitly.

Let $\xi \in G$; let $T^\xi = (T_1^\xi, \cdots, T_n^\xi)$ be the function defined by

$$T_r^\xi(x) = g_r(\xi) + \sum_{s=1}^{n} g_{rs}(\xi)(x_s - \xi_s).$$

T^ξ is thus the linear first approximation to g at ξ; in view of the assumption about $J(x)$, it is a 1-1 map of R^n on to itself.

LEMMA 5.1a. $d(T^\xi(x), g(x))/d(x, \xi) \to 0$ as $x \to \xi$, uniformly on any bounded subset of G.

77

Proof. Since $d(x, y) = || x - y || = (\sum_{r=1}^{n} (x_r - y_r)^2)^{1/2}$, it is enough to show that, for each r,

$$(T_r^\xi(x) - g_r(x))/d(x, \xi) \to 0 \quad \text{as} \quad x \to \xi,$$

uniformly for $\xi \in H$, where H is a given bounded subset of G. Using the first mean-value theorem of differential calculus, we have for some θ between zero and one

$$| T_r^\xi(x) - g_r(x) | = | \sum_{s=1}^{n} g_{rs}(\xi) - g_{rs}(\xi + \theta(x - \xi))) (x_s - \xi_s) |$$

$$\leq (\sum (x_s - \xi_s)^2)^{1/2} (\sum (g_{rs}(\xi) - g_{rs}(\xi + \theta(x - \xi)))^2)^{1/2},$$

so that

$$| T_r^\xi(x) - g_r(x) |/d(x, \xi) \leq (\sum (g_{rs}(\xi) - g_{rs}(\xi + \theta(x - \xi)))^2)^{1/2}.$$

The right side of this last inequality tends to zero as $x \to \xi$, uniformly for $\xi \in H$, in view of the uniform continuity of the functions g_{rs}.

Let I be a given interval, and $\epsilon > 0$. Let I_1, I_2, I_3 be intervals similar to I, with center ξ, and sides $(1 - \epsilon) h$, h, $(1 + \epsilon) h$ times the sides of I, respectively (Fig. 2).

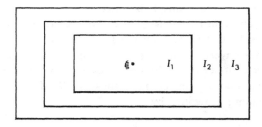

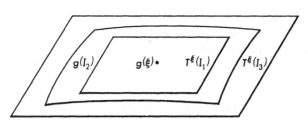

Fig. 2

LEMMA 5.1b. For each bounded subset H of G there is a number $h_0(H)$ such that, if $h < h_0(H)$, and $I_2 \subset G$,

$$T^\xi(I_1) \subset g(I_2) \subset T^\xi(I_3),$$

uniformly for $\xi \in H$.

Proof. $(T^\xi)^{-1}$ exists for each $\xi \in \text{cl } H$, and the elements of $(T^\xi)^{-1}$ are continuous functions of ξ; hence, by Theorem 1.5c, bounded in H. Using the fact that $x = (T^\xi)^{-1} T^\xi(x)$, we thus have

$$d(x, y) \leq K d(T^\xi(x), T^\xi(y))$$

for all x, y and all $\xi \in H$, where K is a constant depending on H. By Lemma 5.1a, $d(T^\xi(x), g(x))/d(x, \xi) \to 0$ as $x \to \xi$. It follows that

$$d(T^\xi(x), g(x))/d(T^\xi(x), g(\xi)) \to 0 \quad \text{as} \quad x \to \xi.$$

(Recall that $T^\xi(\xi) = g(\xi)$.) If now $x \in \Delta I_2$ (the boundary of I_2), we have

$$d(T^\xi(x), g(x))/d(T^\xi(x), g(\xi)) \to 0 \quad \text{as} \quad h \to 0,$$

uniformly for $\xi \in H$. This implies (see Fig. 2) that

$$g(\Delta I_2) \subset T^\xi(I_3) \setminus \text{cl } T^\xi(I_1),$$

uniformly for $\xi \in H$, if h is less than some $h_0(H)$.

It follows at once that $g(I_2) \subset T^\xi(I_3)$. To prove that $T^\xi(I_1) \subset g(I_2)$, we remark first that $g(I_2)$ has a nonempty interior and contains $g(\xi) = T^\xi(\xi)$, and hence contains $T^\xi(I')$ for some interval I' similar to I, with center ξ. Suppose that I' has sides h' times those of I; let $h'' = \sup h'$ for all such h'. Then cl $T^\xi(I'')$ contains a point of $g(\Delta I_2)$, and so $h'' > (1 - \epsilon)h$. The required result is now immediate.

We now deal with the special case where g is linear: $g(x) = Tx$ where T is an $n \times n$ matrix (t_{ij}). Here $g_{ij}(x) = t_{ij}$ and $J(x) = |\det T|$. There are three types of linear transformation—the elementary linear transformations—which are relevant here:

(1) multiplication of a coordinate by a constant:

$$U(x_1, \cdots, x_r, \cdots, x_n) = (x_1, \cdots, kx_r, \cdots, x_n).$$

(2) interchange of two coordinates:

$$V(x_1, \cdots, x_r, \cdots, x_s, \cdots, x_n) = (x_1, \cdots, x_s, \cdots, x_r, \cdots, x_n).$$

(3) replacement of a coordinate by the sum of itself and another:

$$W(x_1, \cdots, x_r, \cdots, x_s, \cdots, x_n) = (x_1, \cdots, x_r + x_s, \cdots, x_s, \cdots, x_n).$$

LEMMA 5.1c. *Any nonsingular linear transformation of R^n to itself can be written as a product of elementary transformations.*

Proof. See any reputable text on matrix theory; for example, L. Mirsky, *Linear Algebra* (Oxford, 1955), Theorem 6.3.1.

LEMMA 5.1d. If E is measurable and T is linear, then $T(E)$ is measurable and

$$m(T(E)) = |\det T| \, m(E).$$

Proof. In view of Lemma 5.1c and the fact that $\det(\Pi_r T_r) = \Pi_r(\det T_r)$, it is enough to prove the lemma when T is an elementary transformation. It is also clear that it is enough to consider the case where E is a bounded interval; the general case can be deduced at once from this.

Since, for transformations of the types U and V above, $U(I)$ and $V(I)$ are intervals when I is an interval, the measurability of $U(I)$ and $V(I)$ and the relation between $m(T(I))$ and $m(I)$ are immediate. In the case of W, things are not quite so simple, since $W(I)$ is not an interval. It is, however, the product of a parallelogram and an $(n-2)$ dimensional interval. Let I be the interval $\{x: a_r < x_r < b_r, 1 \le r \le n\}$ and let N be a positive integer. Write $h = (b_s - a_s)/N$, and let I_k' be the interval (Fig. 3)

$$\{y: a_t \le y_t \le b_t, t \ne r, s; a_s + (k-1)h \le y_s \le a_s + kh;$$

$$a_r + a_s + (k-1)h \le y_r \le b_r + a_s + kh\}.$$

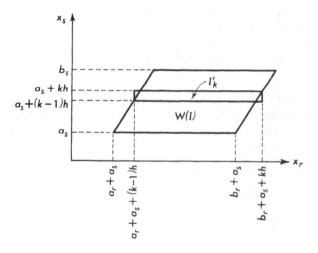

Fig. 3

Then $\bigcup_{k=1}^{N} I_k'$ covers $W(I)$, so that

$$m^*(W(I)) \le \sum m(I_k') = (1 + 1/N) \, m(I).$$

Since N is arbitrary, $m^*(W(I)) \le m(I)$. Similarly, taking inner approximations, $m_*(W(I)) \ge m(I)$, from which the result follows.

COROLLARY. If T is an affine transformation, then if E is measurable so is $T(E)$ and

$$m(T(E)) = Jm(E)$$

where J is the Jacobian of T.

Proof. $T(x) = x_0 + T^*(x)$, where T^* is linear; the Jacobians of T and T^* are the same.

LEMMA 5.1e. If E is a bounded measurable subset of G, then $g(E)$ is measurable and

$$m(g(E)) = \int_E J(x)\, dx.$$

Proof. Take first the case of an interval I; dissect I into N equal subintervals $I_1, \cdots, I_N$ similar to I, with centers $\xi^1, \cdots, \xi^N$, respectively. Let I_r', I_r'' be concentric with I_r and similar to it, with sides $(1 - \epsilon)$, $(1 + \epsilon)$ times the sides of I_r, respectively. By Lemma 5.1b, if N is large enough, then for each ξ^r

$$T^{\xi^r}(I_r') \subset g(I_r) \subset T^{\xi^r}(I_r'').$$

Hence

$$m(T^{\xi^r}(I_r')) \leq m_*(g(I_r)) \leq m^*(g(I_r)) \leq m(T^{\xi^r}(I_r''));$$

that is,

$$(1 - \epsilon)^n\, J(\xi^r)\, m(I_r) \leq m_*(g(I_r)) \leq m^*(g(I_r)) \leq (1 + \epsilon)^n\, J(\xi^r)\, m(I_r).$$

Summing over r,

$$(1 - \epsilon)^n \sum_r J(\xi^r)\, m(I_r) \leq m_*(g(I)) \leq m^*(g(I)) \leq (1 + \epsilon)^n \sum_r J(\xi^r)\, m(I_r),$$

and, since ϵ is arbitrary, it follows that $g(I)$ is measurable.

Since J is continuous, it is certainly integrable. Considering the dissection $\mathscr{E} = \{I_1, \cdots, I_N\}$ of I, we have, for any $\epsilon' > 0$

$$(1 - \epsilon') \sum_r J(\xi^r)\, m(I_r) \leq s_{\mathscr{E}}(J) \leq \int J \leq S_{\mathscr{E}}(J) \leq (1 + \epsilon') \sum_r J(\xi^r)\, m(I_r)$$

provided that N is large enough; and it follows that

$$m(g(I)) = \int_I J(x)\, dx.$$

The extension from an interval I to a general bounded measurable set E is now easy. If $E \in \mathscr{J}$, say $E = \bigcup E_r$, where the E_r are disjoint intervals, then, by Theorem 2.3g,

$$m(g(E)) = \sum m(g(E_r)) = \sum \int_{E_r} J(x)\, dx = \int_E J(x)\, dx,$$

by Lemma 3.2d and Theorem 3.2i, Corollary. If E is the complement of a set in $\mathscr{J}$, the result is now immediate. In general, let E' and E'' be such that $E' \supset E \supset E''$, $E' \in \mathscr{J}$, $\mathscr{C}E'' \in \mathscr{J}$, $m(E' \setminus E'') < \epsilon$. Then

$$m^*(g(E)) \leq m(g(E')) = \int_{E'} J(x)\, dx,$$

$$m_*(g(E)) \geq m(g(E'')) = \int_{E''} J(x)\, dx,$$

and an application of Theorem 3.2i now gives the required result.

THEOREM 5.1f. Let $E \subset G$ and $F = g(E)$. Then $f(y)$ is integrable over F if and only if $f(g(x)) J(x)$ is integrable over E; and, if so,

$$\int_F f(y)\, dy = \int_E f(g(x)) J(x)\, dx.$$

Proof. It is sufficient to prove that if $f(y)$ is integrable over F then $f(g(x)) J(x)$ is integrable over E, and $\int_F f(y)\, dy = \int_E f(g(x)) J(x)\, dx$. For then, in view of the relation

$$| \det \partial y_j / \partial x_i | \, | \det \partial x_j / \partial y_i | = 1,$$

if $y = g(x)$, it will follow that if $f(g(x)) J(x)$ is integrable over E then $f(g(g^{-1}(y))) J(x) J(y) = f(y)$ is integrable over F, and the two integrals are equal.

Suppose first that f is a step function; in this case the result is an immediate consequence of Lemma 5.1e.

If f is a bounded function and E is a bounded set, there are step functions ϕ_1, ϕ_2 such that $\phi_1 \geq f \geq \phi_2$ and $\int_E \{\phi_1(y) - \phi_2(y)\}\, dy < \epsilon$. Then

$$\int_E \{\phi_1(g(x)) - \phi_2(g(x))\} J(x)\, dx < \epsilon,$$

so that

$$\int_E^* f(g(x)) J(x)\, dx - \int_{*E} f(g(x)) J(x)\, dx < \epsilon$$

and

$$\left| \int_E^* f(g(x)) J(x)\, dx - \int_F f(y)\, dy \right| < \epsilon.$$

From this it follows—ϵ being arbitrary—that $\int_E f(g(x)) J(x)\, dx$ exists and is equal to $\int_F f(y)\, dy$.

If f and E are not necessarily bounded, the result follows from Lemma 4.1c, on writing f as the sum of its positive and negative parts.

5.2. Differentiation of Integrals

The relation

$$\frac{d}{dx}\int_a^x f(t)\,dt = f(x)$$

—under suitable conditions—is familiar in elementary calculus. It is studied here in the context of the Lebesgue integral. Attention is restricted to functions of a single variable. So far, of course, the symbol $\int_a^b f(t)\,dt$ has not been defined; we make the usual definition,

$$\int_a^b f(t)\,dt = \int_{(a,b)} f(t)\,dt \qquad (b \geq a),$$
$$= -\int_{(b,a)} f(t)\,dt \quad (b < a).$$

For completeness, we include a proof of the following elementary result:

LEMMA 5.2a. If f is continuous at x, then $d(\int_a^x f(t)\,dt)/dx$ exists and is equal to $f(x)$.

Proof. Write $F(x) = \int_a^x f(t)\,dt$; then

$$\left| \frac{F(x+h) - F(x)}{h} - f(x) \right| = \left| \frac{1}{h}\int_x^{x+h} \{f(t) - f(x)\}\,dt \right|$$
$$\leq \frac{1}{|h|}\int_x^{x+h} |f(t) - f(x)|\,dt.$$

Given ϵ, there exists δ such that $|f(t) - f(x)| < \epsilon$ whenever $|t - x| < \delta$, by the assumed continuity of f at x. It follows that the left side of the above inequality is less than ϵ whenever $|h| < \delta$, which is what is wanted.

In order to study differentiability in general, it is convenient to introduce the four derivatives

$$D^+f(x) = \limsup_{h\to 0+} \{f(x+h) - f(x)\}/h$$

$$D_+f(x) = \liminf_{h\to 0+} \{f(x+h) - f(x)\}/h$$

$$D^-f(x) = \limsup_{h\to 0-} \{f(x+h) - f(x)\}/h$$

$$D_-f(x) = \liminf_{h\to 0-} \{f(x+h) - f(x)\}/h.$$

These are, respectively, the *right upper, right lower, left upper,* and *left lower derivatives* of f at x. Their existence (as finite real numbers or $+\infty$ or $-\infty$) is guaranteed at all points where the function f is finite. It is clear that a necessary and sufficient condition that $Df(x)$ ($\equiv df(x)/dx$) should exist is that the four derivatives should be equal.

The next result is a covering theorem of the same general type as the Heine-Borel theorem (Theorem 1.4a) but essentially more complicated. There is also an n dimensional form of the theorem, which is no more difficult than the one dimensional form (which is stated and proved below).

Let A be a subset of R, $\mathscr{I}$ a collection of intervals such that each point of A is contained in an arbitrarily small interval of $\mathscr{I}$. Then $\mathscr{I}$ is a *Vitali covering* of A. ($\mathscr{I}$ covers A in the sense of Vitali.)

THEOREM 5.2b (Vitali). Let $\mathscr{I}$ be a Vitali covering of A by nondegenerate intervals; then there is a countable disjoint subset $I_1, I_2, \cdots$ of $\mathscr{I}$ such that

$$m(A \setminus \bigcup I_r) = 0.$$

Proof. There is no loss of generality in supposing that the intervals of $\mathscr{I}$ are closed, for, if each interval I is replaced by its closure cl I, we have again a Vitali covering. If the theorem has been proved for coverings by closed intervals, then from $m(A \setminus \bigcup \operatorname{cl} I_r) = 0$ it follows that $m(A \setminus \bigcup I_r) = 0$, which proves the general result.

Suppose first that A is a subset of some bounded open set G. Then it may be assumed that each interval of $\mathscr{I}$ is also in G, for the intervals of $\mathscr{I}$ which are subsets of G evidently constitute a Vitali covering of A.

Let I_1 be chosen arbitrarily. Suppose that $I_1, \cdots, I_s$ have been chosen; let $k_s = \sup m(I)$, over all $I \in \mathscr{I}$ such that $I \cap I_r = \varnothing$ for $1 \le r \le s$. If $I_1, \cdots, I_s$ cover A, the theorem is trivial; otherwise, $k_s > 0$ and there is an interval I_{s+1}, disjoint from $I_1, \cdots, I_s$, with $m(I_{s+1}) > k_s/2$.

If the process does not terminate (if it does, the theorem is immediate), we can choose an infinite sequence of intervals I_r in this way. For any $I \in \mathscr{I}$, there is an interval I_r of the sequence with $I \cap I_r \ne \varnothing$. For, if not, $k_r \ge m(I)$ for all r, and so $m(I_r) \ge m(I)/2$ for all r. Since $\sum m(I_r)$ is convergent (the intervals being disjoint and contained in G), it follows that $m(I_r) \to 0$; hence $m(I) = 0$, which is not possible, since the intervals of $\mathscr{I}$ are nondegenerate.

To prove $m(A \setminus \bigcup I_r) = 0$, suppose that $m^*(A \setminus \bigcup I_r) > 0$. Denote by I_r' the interval concentric with I_r, and of five times the length. Since $\sum m(I_r)$ is convergent, so is $\sum m(I_r')$; there exists N such that

$$\sum_{r=N+1}^{\infty} m(I_r') < m^*(A \setminus \bigcup I_r).$$

It follows that there exists $x \in A \setminus \bigcup I_r$, such that $x \notin I_r'$ for $r > N$; for, if not, $\bigcup_{r=N+1}^{\infty} I_r'$ would cover $A \setminus \bigcup I_r$ and we would have

$$m^*(A \setminus \bigcup I_r) \le m\left(\bigcup_{r=N+1}^{\infty} I_r' \right) \le \sum_{r=N+1}^{\infty} m(I_r'),$$

contradicting our choice of N.

By the definition of a Vitali covering, the chosen point x is in some $I \in \mathscr{I}$. Since $I_1 \cup \cdots \cup I_N$ is closed and $x \notin I_1 \cup \cdots \cup I_N$, we can assume that this I is disjoint from $I_1, \cdots, I_N$. Let M be the least integer r such that $I \cap I_r \neq \varnothing$; then $M > N$, so that $x \notin I'_M$. We now have $x \in I, I \cap I_M \neq \varnothing, x \notin I'_M$, and it follows that $m(I) > 2m(I_M) > k_{M-1}$. But also, since $I \cap I_r = \varnothing$ for $r = 1, 2, \cdots, M - 1$, we have $m(I) \leq k_{M-1}$; this provides a contradiction, and so the theorem is proved for bounded A.

To extend the result to unbounded sets A, consider $A_r = A \cap {]}r, r + 1{[}$ $(-\infty < r < \infty)$. It is clear that a Vitali covering of A is also a Vitali covering of each A_r. Apply the result already proved to A_r and note that $A = \bigcup A_r \cup B$, where B is of measure zero; the general result follows at once.

COROLLARY. If $m^*(A) < \infty$ there is a finite set of disjoint intervals I_r such that $m^*(A \setminus \bigcup_{r=1}^{P} I_r) < \epsilon$; (or, equivalently, $m^*(A \cap \bigcup_{r=1}^{P} I_r) > m^*(A) - \epsilon$).

In the next three results f is assumed to be a (finite-valued) monotonic function; without loss of generality it may be assumed to be increasing: $f(x) \geq f(y)$ if $x \geq y$.

LEMMA 5.2c. $m\{x: D^+f(x) = +\infty\} = m\{x: D_+f(x) = -\infty\} = m\{x: D^-f(x) = +\infty\} = m\{x: D_-f(x) = -\infty\} = 0$.

Proof. It is enough to consider the first of the sets specified; the proof for the others is similar. It may evidently be assumed that all sets are contained in some fixed finite interval (a, b). Let $A = \{x: D^+f(x) = +\infty\}$, and suppose $m^*(A) = k > 0$. Then, for each $x \in A$ there is, for any given $K > 0$, an arbitrarily short interval $[x, x + h]$ such that $f(x + h) - f(x) > Kh$. By Theorem 5.2b, Corollary, there is a finite disjoint set of such intervals, of total measure at least $k/2$. Summing the above inequality over this set of intervals,

$$f(b) - f(a) \geq \sum (f(x + h) - f(x)) \geq Kk/2;$$

and since K is arbitrarily large and the function is everywhere finite, this gives a contradiction. It follows that $k = 0$, as required.

LEMMA 5.2d. $m\{x: D^+f(x) > D_-f(x)\} = m\{x: D^-f(x) > D_+f(x)\} = m\{x: D^+f(x) > D_+f(x)\} = m\{x: D^-f(x) > D_-f(x)\} = 0$.

Proof. As for the previous lemma, only the first set need be considered. Let $A = \{x: D^+f(x) > D_-f(x)\}$; suppose $m^*(A) > 0$. Let r, s be rational numbers such that $r < s$; write $A_{rs} = \{x: D_-f(x) < r < s < D^+f(x)\}$. Then it is clear that $A = \bigcup A_{rs}$. Since $m^*(A) \leq \sum m^*(A_{rs})$, by Lemma 2.3e, it follows that $m^*(A_{rs}) > 0$ for at least one pair r, s. Let A' be any such set A_{rs}, and let $m^*(A') = k > 0$. As before, there is no loss of generality in assuming that all sets are subsets of a bounded interval (a, b). Let G be an open set containing A', with $m(G) < k + \epsilon$; for any $\epsilon > 0$ such a set exists, by Theorem 2.2h.

Each $x \in A'$ is contained in arbitrarily small intervals $[x - h, x]$ $(h > 0)$ such that $f(x) - f(x - h) < rh$, since $D_- f(x) < r$. We may suppose, without loss of generality, that these intervals are all in G; they form a Vitali covering of A'. By Theorem 5.2b, Corollary, there is a finite disjoint set $I_1, \cdots, I_p$ of them such that $I = I_1 \cup I_2 \cup \cdots \cup I_p$ covers a subset of A' of outer measure greater than $k - \epsilon$. Summing over these, we have

$$\sum_{t=1}^{p} (f(x_t) - f(x_t - h_t)) < r \sum_{t=1}^{p} h_t = rm(I) < rm(G) < r(k + \epsilon).$$

Now let $A'' = A' \cap \text{int } I$. Each $x \in A''$ is in arbitrarily small intervals $[x, x + h]$ $(h > 0)$ such that $f(x + h) - f(x) > sh$; and it may be assumed that all such intervals are contained in I. As before, there is a finite disjoint set of them, $I_1', \cdots, I_q'$ such that $m(I_1' \cup \cdots \cup I_q') > m^*(A'') - \epsilon > k - 2\epsilon$. Summing over these intervals,

$$\sum_{t=1}^{q} (f(x_t' + h_t') - f(x_t')) > s \sum_{t=1}^{q} h_t' > s(k - 2\epsilon).$$

Since each interval I_t' is contained in some I_u, it follows that (since f is monotonic increasing)

$$\sum_{t=1}^{q} (f(x_t' + h_t') - f(x_t')) \le \sum_{u=1}^{p} (f(x_u) - f(x_u + h_u)),$$

and hence that

$$r(k + \epsilon) > s(k - 2\epsilon).$$

This gives a contradiction, if $\epsilon < k(s - r)/(r + 2s)$, and so the required result is established.

THEOREM 5.2e. A monotonic function is differentiable almost everywhere.

Proof. Let f be a (finite-valued, increasing) monotonic function; by Lemmas 5.2c and 5.2d we have, almost everywhere,

$$-\infty < D_+ f \le D^+ f \le D_- f \le D^- f \le D_+ f \le D^+ f < \infty.$$

It follows that Df exists (as a finite real number) almost everywhere.

From Theorem 5.2e it can be seen at once (cf. the proof of Theorem 5.2g, below) that $\int_a^x f(t)\,dt$ is differentiable p.p. To show that the derivative is what it should be, we require another lemma:

LEMMA 5.2f. If $\int_a^x g = 0$ for $a \le x \le b$, then $g = 0$ p.p. in $[a, b]$.

Proof. Suppose the conclusion is false; let $g > 0$ in a set E of positive measure. There is a set $J \in \mathscr{J}$ such that $J \supset \mathscr{C}E$, $m(\mathscr{C}J) > 0$. (Complements here are, of course, relative to $[a, b]$.) It is evident that $\int_I g = 0$ for all subintervals I of $[a, b]$; from this it follows (using Theorem 3.2i, Corollary) that $\int_J g = 0$ for all $J \in \mathscr{J}$. Hence, taking complements, $\int_{\mathscr{C}J} g = 0$ for all J. This is inconsistent with g being strictly positive in a set of the form $\mathscr{C}J$ of positive measure (Theorem 3.2h). A similar contradiction arises if g is negative in a set of positive measure, and the result follows.

THEOREM 5.2g. If f is integrable, then $\int_a^x f(t)\, dt$ is differentiable p.p. with respect to x; and the derivative is equal to f p.p.

Proof. Write $F(x) = \int_a^x f(t)\, dt$, and F' for the derivative of F. If $f \ge 0$ then F is monotonic, and so F' exists p.p. by Theorem 5.2e. In general, write $f = f^+ - f^-$, where each of f^+, f^- is nonnegative; then F is the difference of two monotonic functions, and again F' exists p.p.

It remains to prove that $F' = f$ p.p. Suppose first that f is bounded; say $|f| \le K$. Then, if $h \ne 0$,

$$\left| \frac{1}{h}\{F(x+h) - F(x)\} \right| = \left| \frac{1}{h} \int_x^{x+h} f(t)\, dt \right| \le K.$$

Now, as $h \to 0$, the left side of this tends to $F'(x)$ p.p. Let $h_1, h_2, \cdots$ be any sequence of values of h, tending to zero; it follows from Lebesgue's theorem on bounded convergence (Theorem 4.1b, Corollary) that

$$\int_a^x F'(t)\, dt = \lim_{r \to \infty} \frac{1}{h_r} \int_a^x (F(t + h_r) - F(t))\, dt$$

$$= \lim_{h \to 0} \frac{1}{h} \int_a^x (F(t + h) - F(t))\, dt$$

$$= \lim_{h \to 0} \frac{1}{h} \int_x^{x+h} F(t)\, dt - \lim_{h \to 0} \frac{1}{h} \int_a^{a+h} F(t)\, dt.$$

Now, F is a continuous function, by Theorem 3.2i, and so, by Lemma 5.2a,

$$\int_a^x F'(t)\, dt = F(x) - F(a) = F(x) = \int_a^x f(t)\, dt,$$

so that $\int_a^x (F'(t) - f(t))\, dt = 0$; and it follows by Lemma 5.2f that $F'(t) = f(t)$ p.p.

If f is unbounded, it is evidently sufficient to consider the case $f \ge 0$, as the general case will follow immediately. Let $f^{[k]}(x)$ be the kth truncate of f (Sec. 4.1); by the part of the theorem just proved,

$$\frac{d}{dx} \int_a^x f^{[k]}(t)\, dt = f^{[k]}(x) \text{ p.p.}$$

Also, for each k, $\int_a^x (f(t) - f^{[k]}(t))\, dt$ is an increasing function of x; its derivative exists p.p. by Theorem 5.2e, and is evidently nonnegative..Hence $F' \geq f^{[k]}$p.p.; making $k \to \infty$, it follows that $F' \geq f$ p.p. Moreover, choose a sequence $h_1, h_2, \cdots$ of positive real numbers, tending to zero, such that

$$\liminf_{h \to 0+} \int_a^b \frac{1}{h} (F(t + h) - F(t))\, dt = \liminf_{r \to \infty} \int_a^b \frac{1}{h_r} (F(t + h_r) - F(t))\, dt,$$

and apply Theorem 4.1d (Fatou's theorem); we have

$$\int_a^b F'(t)\, dt \leq \liminf_{h \to 0+} \int_a^b \frac{1}{h} (F(t + h) - F(t))\, dt.$$

As in the bounded case, the right side is $\int_a^b f(t)\, dt$, and hence

$$\int_a^b (F'(t) - f(t))\, dt \leq 0.$$

Since it has been established that $F' \geq f$ p.p., Theorem 3.2h can be applied to show that $F' = f$ p.p., as required.

5.3. Integration of Derivatives

In general, the relation

$$\int_a^b f'(x)\, dx = f(b) - f(a) \tag{*}$$

(where f' is the derivative of f) does not hold for Lebesgue integrals. For instance, let $f(x)$ be 0 in $[0, 1/2]$ and 1 in $]1/2, 1]$; then $f'(x) = 0$ p.p. and

$$\int_0^1 f'(x)\, dx = 0 \neq f(1) - f(0).$$

This situation may arise even when the function is continuous; consider, for example, the Cantor function g defined in Sec. 1.2. This function is evidently continuous, and its derivative exists and is zero in $\mathscr{C}T$, that is, p.p. in $[0, 1]$. And

$$\int_0^1 g'(x)\, dx = 0 \neq g(1) - g(0).$$

The results of Sec. 5.2 show that the relation (*) holds when $f(x)$ is of the form $\int_c^x \phi(y)\, dy$, where ϕ is integrable. Indeed, this condition is essentially necessary as well as sufficient. If ϕ is integrable (over some suitable set), we define an *indefinite integral* of ϕ to be a function f of the form

$$f(x) = \alpha + \int_c^x \phi(y)\, dy,$$

where α is a constant. Two remarks should be made. In the first place, an indefinite integral is not, in general, of the form $\int_c^x \phi(y)\, dy$ for suitably chosen c;

consider, for example, $f(x) \equiv 1$. Secondly, the condition that $f'(x) = \phi(x)$ p.p. is not sufficient to ensure that f is an indefinite integral of ϕ. For example, the function g referred to in the previous paragraph is not an indefinite integral of 0.

THEOREM 5.3a. A necessary and sufficient condition that

$$\int_a^b f'(y)\, dy = f(b) - f(a)$$

(for all relevant a, b) is that f should be an indefinite integral.

Proof. The sufficiency is clear, in view of Theorem 5.2g. The necessity follows at once on taking $b = x$.

In Chap. 6 we discuss briefly another condition (absolute continuity) for a function to be an indefinite integral.

5.4. Integration by Parts

The well-known formula is established in the following form:

THEOREM 5.4a. Let F, G be indefinite integrals of f, g, respectively; then

$$F(x)\, G(x) - F(c)\, G(c) = \int_c^x \{f(y)\, G(y) + F(y)\, g(y)\}\, dy.$$

Proof. Suppose first that f and g are continuous. The functions F and G are continuous in any case (Theorem 3.2i), and it follows by Lemma 5.2a that the function

$$F(x)\, G(x) - \int_c^x \{f(y)\, G(y) + F(y)\, g(y)\}\, dy$$

has a derivative which is everywhere zero. By the first mean-value theorem of differential calculus, the function is a constant, which is obviously $F(c)\, G(c)$.

If f and g are not necessarily continuous, let f_r and g_r be continuous approximations to f and g in the sense of Theorem 4.3b, so that

$$\int_c^x |f - f_r| \to 0, \qquad \int_c^x |g - g_r| \to 0$$

as $r \to \infty$. Then—with the obvious notation—$F_r \to F$ and $G_r \to G$, uniformly for $c \le x \le d$ (say). It follows that

$$F(x)\, G(x) - F(c)\, G(c) = \lim \{F_r(x)\, G_r(x) - F_r(c)\, G_r(c)\}$$

$$= \lim \int_c^x \{f_r(y)\, G_r(y) + F_r(y)\, g_r(y)\}\, dy$$

$$= \int_c^x \{f(y)\, G(y) + F(y)\, g(y)\}\, dy,$$

as required.

EXERCISES

1. Justify the equation

$$\int_0^\infty \int_0^\infty f(x, y) \, dx \, dy = \int_0^{\pi/2} \int_0^\infty g(r, \theta) \, r \, dr \, d\theta,$$

where $x = r \cos \theta$, $y = r \sin \theta$, $g(r, \theta) = f(r \cos \theta, r \sin \theta)$, and $f(x, y)$ is integrable over the quadrant $0 \le x$, $0 \le y$.

2. Examine in detail the operations

$$\int_0^\infty e^{-x^2} \, dx \int_0^\infty e^{-y^2} \, dy = \int_0^\infty \int_0^\infty e^{-x^2 - y^2} \, dx \, dy$$

$$= \int_0^{\pi/2} \int_0^\infty e^{-r^2} r \, dr \, d\theta$$

$$= \pi/4$$

involved in proving that $\int_0^\infty e^{-x^2} \, dx = (\sqrt{\pi})/2$. (The result can, of course, be established quite rigorously without any reference to the Lebesgue integral.)

3. Show that the conclusion of Theorem 5.1f is still true, even if $J(x)$ is sometimes zero, provided that, given $\epsilon > 0$, there exists $\delta > 0$ such that

$$m \left(\{x \colon J(x) \le \delta\} \right) < \epsilon.$$

4. A one-parameter family of transformations

$$y = g^t(x)$$

is defined in R^{2n}, such that $g^0(x) = x$, and

$$\frac{\partial y_r}{\partial t} = - \frac{\partial H}{\partial y_{n+r}}, \qquad \frac{\partial y_{n+r}}{\partial t} = + \frac{\partial H}{\partial y_r} \qquad (1 \le r \le n),$$

where H is a function of y and t. By considering $\partial J'/\partial t$, where $J' = \det \partial y_r / \partial x_s$, prove (making any convenient continuity and differentiability assumptions) that $m(g^t(E)) = m(E)$ for all measurable sets E and all t. (This is *Liouville's theorem* on Hamiltonian dynamical systems.)

5. Defining $SDf(x) = \lim\limits_{h \to 0} \dfrac{f(x + h) - f(x - h)}{2h}$, prove that if f is an increasing function then $SD \int_a^x f(y) \, dy$ exists for all x. How is it related to f at the points where f is discontinuous?

6. Show that $SD \int_a^x f(y) \, dy$ exists for all $x \in [a, b]$ if f is of bounded variation in $[a, b]$. (See Sec. 6.4.) Produce an example to show that $SD \int_a^x f(y) \, dy$ need not exist at all points $x \in [a, b]$ even though f is integrable over $[a, b]$.

7. Give an example of a function for which the four derivatives are all different at $x = 0$.

8. Prove that the four derivatives of a (finite-valued) measurable function are measurable.

9. Show that Vitali's theorem (Theorem 5.2b), as stated in dimension $n = 1$, is false if $n \geq 2$. Prove that it is true if the intervals of the covering are of restricted shape, in the sense that the ratio of the greatest side to the least side is bounded. (In particular, the theorem will hold for squares, cubes, $\cdots$.)

10. Give an example of a monotonic function on $[0, 1]$ which is not differentiable at an uncountable set of points.

11. Let f be any function; show that the set of points at which

$$\pm \infty \neq D^+f = D_+f \neq D^-f = D_-f \neq \pm \infty$$

(that is, at which the right and left derivatives exist and are finite but unequal) is countable.

12. Prove that if f is a monotonic increasing function, then

$$\int_a^b f'(x) \, dx \leq f(b) - f(a).$$

[6]

More General Measures

6.1. Borel Measures

In many situations, properties of a set other than its size are of interest. For example, in elementary physics we may require to know (for a nonuniform mass distribution) the mass of material associated with a set in R^3. This is not proportional to the Lebesgue measure of the set, but one feels that it should have similar properties. Also, we may consider the "probability" (under given circumstances) that a point should be in a subset of R^n; again, similar properties are to be expected. A third example—slightly more complicated—is that of a distribution of electric charges in R^3. Here the total charge concentrated on a set is of interest, and this may be positive or negative. (In the first two cases the mass and probability were essentially nonnegative.) For the moment we leave this generalization aside, returning to it in Sec. 6.2. We proceed to formulate a definition of measure suggested by the examples just quoted and by the properties of Lebesgue measure obtained in earlier chapters.

Let X be any set, and $\mathscr{F}$ a σ ring (Sec. 2.5) of subsets of X. An extended-real function μ is a *measure* on $\mathscr{F}$ (less correctly, a measure on X) if

(1) $\mu(F) \geq 0$ for all $F \in \mathscr{F}$;
(2) $\mu(\varnothing) = 0$;
(3) if $F \in \mathscr{F}$ is a countable disjoint union $\bigcup F_r$, where $F_r \in \mathscr{F}$ for all r,

then

$$\mu(F) = \sum \mu(F_r).$$

Measures may be defined on classes of sets more general than σ rings, but we ignore this generalization here. The sets of $\mathscr{F}$ are *measurable with respect to* μ (μ measurable).

EXAMPLES. (1) Let X be any set, $\mathscr{F}$ the class of all subsets of X. For each $x \in X$, let $w(x)$ be a nonnegative real number. Then

$$\mu(F) = \sum_{x \in F} w(x)$$

defines a measure on $\mathscr{F}$; $\mu(F)$ is finite if and only if $w(x) = 0$ for all but a countable subset $\{x_r\}$ of F, and $\sum w(x_r)$ is convergent.

92

(2) Let X be R^n, $\mathscr{F}$ the Lebesgue-measurable sets in R^n, and μ Lebesgue measure.

(3) Let X be R^n, $\mathscr{F}$ the Borel sets in R^n, and μ Lebesgue measure.

(4) Let X be a bounded Lebesgue-measurable subset of R^n, $\mathscr{F}$ the measurable subsets of X, and f a nonnegative measurable function. Then

$$\mu(F) = \int_F f(x)\,dx \quad \text{if } f \text{ is integrable over } F,$$

$$= \infty \qquad \text{if } f \text{ is not integrable over } F$$

defines a measure on $\mathscr{F}$, by Theorem 3.2i.

It is immediate that we can define integrability and integrals with respect to a measure μ, in just the same way as for Lebesgue measure m (Chap. 3). The integral of f over E with respect to μ is denoted $\int_E f\,d\mu$ or $\int_E f(x)\,d\mu(x)$. Measurability is now measurability with respect to μ (μ measurability). "Almost everywhere" is now "almost everywhere with respect to μ," that is, everywhere except on a set F such that $\mu(F) = 0$; this is usually written "p.p. $[\mu]$." All the considerations of Secs. 3.1 to 3.4 remain valid, with the possible exception of those involving continuous functions (Theorem 3.3b and Corollary 2 of Theorem 3.3e). Sec. 4.1, Sec. 4.2 (if one makes the appropriate definition of product measure), and parts of Secs. 4.3, 4.4, and 4.5 continue to hold, with the appropriate changes.

The measure μ is *finite* if $\mu(X)$ is finite; it is σ *finite* if X is a countable union of sets of finite μ measure.

From now on we restrict attention to subsets of R^n. We shall require our measures to be defined on (at least) the Borel sets $\mathscr{E}$ of R^n, or, more generally, of some Borel subset X of R^n. This is equivalent to the requirement that $X \cap I$ shall be μ measurable for every interval I. Moreover, we require that $\mu(E)$ be finite if E is a compact Borel set. Under these conditions, μ is a *Borel measure* on $\mathscr{E}$.

Strictly, one ought to distinguish between measures such as (2) and (3) above, which are defined on different σ rings; but it should cause no confusion if they are identified. Thus Lebesgue measure m will be regarded as a Borel measure. If, in example (1) above, $X = R^n$ and $\sum_{x \in E} w(x) < \infty$ whenever E is bounded, the measure μ is a Borel measure. In example (4), μ is a Borel measure if f is integrable over every compact subset of X.

For the rest of this chapter, "measure" will mean "Borel measure," and "measurable set" will mean "Borel set." Some results are, of course, valid in much more general situations.

In distinction to Lebesgue measure, it is *not*, in general, true that two intervals with the same end points (such as $[a, b]$ and $]a, b[$) necessarily have the same measure. To take a trivial example, if δ is the measure on R defined by $\delta(E) = 1$ if $0 \in E$, $\delta(E) = 0$ if $0 \notin E$, then $\delta([0, 1]) = 1$ whereas $\delta(]0, 1[) = 0$.

If μ_1 and μ_2 are measures, their sum $\mu_1 + \mu_2$, defined by

$$(\mu_1 + \mu_2)(E) = \mu_1(E) + \mu_2(E)$$

for all $E \in \mathscr{E}$, is again a measure, as is trivially verifiable. Again, if c is a non-negative real number and μ is a measure, then $c\mu$, defined by

$$(c\mu)(E) = c\mu(E)$$

for $E \in \mathscr{E}$, is also a measure. The zero measure (for which the measure of every Borel set is zero) is denoted by 0.

For Borel measures, the analogue of Theorem 3.3b is valid, together with its consequences:

THEOREM 6.1a. A continuous function is measurable and integrable over every compact Borel set. Every continuous function of compact support is integrable over R^n with respect to every Borel measure on R^n. Every bounded continuous function is integrable over R^n with respect to every finite measure.

Proof. It is assumed that the domain of definition of the function is a Borel set. Then, just as in Theorem 3.3b, the function is measurable because every open set is a Borel set (Theorem 2.2g). The rest of the theorem now follows from Theorem 3.3e (or rather, its extension to general Borel measures).

What distinguishes Lebesgue measure among all the other measures on R^n? There is a very easy answer. We say that a measure μ is *translation invariant* if $\mu(E - x) = \mu(E)$ for all $x \in R^n$, $E \in \mathscr{E}$. (Recall from Sec. 1.3 that $E - x = \{y : y = z - x, \ z \in E\}$.) It is evident that Lebesgue measure is translation invariant. We proceed to prove that this property is characteristic.

THEOREM 6.1b. A measure μ is translation invariant if and only if it is a constant multiple of Lebesgue measure m; that is, if there is a constant k such that

$$\mu(E) = km(E)$$

for all Borel sets E.

Proof. It is sufficient to prove the required relation for bounded intervals; the general case then follows by the countable additivity of μ and m. Let $\mathscr{S}$ be the class of half-open intervals with rational sides, that is, intervals of the form

$$I = \{x : a_r \leq x_r < b_r, \ 1 \leq r \leq n\},$$

where $b_r - a_r$ is rational for $1 \leq r \leq n$. Let

$$I_N = \{x : 0 \leq x_r < N^{-1}, \ 1 \leq r \leq n\}.$$

Then, for a suitable integer N, we have

$$b_r - a_r = p_r N^{-1} \quad (1 \leq r \leq n),$$

where $p_1, \cdots, p_n$ are integers. If then μ is translation invariant, we have

$$\mu(I) = p_1 p_2 \cdots p_n \mu(I_N) = \frac{m(I)}{m(I_N)} \mu(I_N),$$

so that

$$\frac{\mu(I)}{m(I)} = \frac{\mu(I_N)}{m(I_N)}.$$

It follows that if $I, I' \in \mathscr{S}$ we have

$$\frac{\mu(I)}{m(I)} = \frac{\mu(I')}{m(I')}.$$

Let this ratio be k; then $\mu(I) = km(I)$ for $I \in \mathscr{S}$.

Suppose now that I is a general bounded interval; given $\epsilon > 0$, let I', $I'' \in \mathscr{S}$ be such that $I' \supset I \supset I''$, and

$$m(I') < (1 + \epsilon) m(I), \qquad m(I'') > (1 - \epsilon) m(I).$$

Then

$$\mu(I) \leq \mu(I') = km(I') < k(1 + \epsilon) m(I),$$
$$\mu(I) \geq \mu(I'') = km(I'') > k(1 - \epsilon) m(I),$$

so that—ϵ being arbitrary—$\mu(I) = km(I)$, as required.

6.2. Signed Measures and Complex Measures

In order to distinguish it from the more general measures which follow, a measure as defined in Sec. 6.1 will now be called a *positive measure* (more accurately, a nonnegative measure). A function μ, defined on the Borel sets $\mathscr{E}$, is a *signed* (Borel) *measure* if it is of the form

$$\mu(E) = \mu_1(E) - \mu_2(E) \qquad (E \in \mathscr{E}),$$

where μ_1, μ_2 are positive (Borel) measures, at least one of which is finite. This last restriction excludes difficulties which would arise in defining $\mu(E)$ if we had $\mu_1(E) = \mu_2(E) = \infty$. The measure μ is finite if both μ_1 and μ_2 are finite. A signed measure may also be called a *real measure*, to distinguish it from a *complex measure*, which is a set function of the form

$$\mu(E) = \mu_1(E) + i\mu_2(E) \qquad (E \in \mathscr{E}),$$

where μ_1 and μ_2 are signed measures. A complex measure if finite of both its real and imaginary parts, μ_1 and μ_2, are finite. A complex measure is a linear combination of at most four positive measures. Integrals with respect to signed measures and complex measures are defined in the obvious way in terms of integrals with respect to the appropriate positive measures. For example, if $\mu = \mu_1 - \mu_2 + i(\mu_3 - \mu_4)$, then f is said to be integrable with respect to μ if and only if it is integrable with respect to μ_r for $1 \leq r \leq 4$; and then

$$\int f\, d\mu = \int f\, d\mu_1 - \int f\, d\mu_2 + i \int f\, d\mu_3 - i \int f\, d\mu_4.$$

A (real) constant multiple of a signed measure is a signed measure. The sum or difference of two measures, however, is not, in general, a measure, and so the signed measures do not form a linear space. The finite signed measures evidently form a real linear space, and the finite complex measures a complex linear space. Some of the techniques of linear-space theory can be applied with advantage to certain problems involving measures.

LEMMA 6.2a. If μ is a real or complex measure, then $| \mu |$, defined by

$$| \mu |\, (E) = \sup \sum | \mu(E_r') |,$$

the supremum being over all finite disjoint unions $E = \bigcup E_r'$, is a positive measure.

Proof. The first two of the three defining properties of a measure (Sec. 6.1) are obvious. It remains to verify the third. Suppose that E is written as a countable disjoint union $\bigcup E_s$. Evidently

$$| \mu |\, (E) \geq \sum_{s=1}^{N} | \mu |\, (E_s)$$

for all N, and so if $\sum | \mu |\, (E_s)$ diverges, then $| \mu |\, (E) = \infty$. If $\sum | \mu |\, (E_s) < \infty$, it is clear that $| \mu |\, (E) \geq \sum | \mu |\, (E_s)$, and we now have to establish the opposite inequality. Let $\epsilon > 0$ be given; let $E = \bigcup E_r'$ be a finite disjoint decomposition of E such that

$$| \mu |\, (E) < \sum | \mu\, (E_r') | + \epsilon.$$

Write $E_{rs} = E_r' \cap E_s$; then $\mu(E_r') = \sum_s \mu(E_{rs})$ and $| \sum_s \mu(E_{rs}) | \leq \sum_s | \mu(E_{rs}) |$, whence

$$| \mu |\, (E) < \sum_s \sum_r | \mu(E_{rs}) | + \epsilon \leq \sum_s | \mu |\, (E_s) + \epsilon,$$

which proves that $| \mu |$ has the required additivity property.

To complete the verification that $|\mu|$ is a Borel measure, it has to be shown that $|\mu|(E) < \infty$ if E is compact. This, however, is obvious, since if μ is a signed measure, $\mu = \mu_1 - \mu_2$,

$$|\mu|(E) \leq \sum |\mu(E_r')| + \epsilon \quad \text{(for suitable } E_r')$$
$$\leq \sum (\mu_1(E_r') + \mu_2(E_r')) + \epsilon$$
$$= \mu_1(E) + \mu_2(E) + \epsilon,$$

with a similar inequality in the complex case.

The measure $|\mu|$ is the *total variation* of μ. If μ is real, then $(|\mu| + \mu)/2$, $(|\mu| - \mu)/2$ are obviously positive measures. They are, respectively, the *positive variation* (upper variation) μ^+ and the *negative variation* (lower variation) μ^- of μ. We thus have

$$\mu = \mu^+ - \mu^-, \qquad |\mu| = \mu^+ + \mu^-.$$

The expression of μ as the difference of μ^+ and μ^- is the *Jordan decomposition* of μ.

The expression of μ as a difference of positive measures is never unique, for if μ' is any finite positive measure, then $\mu = (\mu^+ + \mu') - (\mu^- + \mu')$. The Jordan decomposition, however, is in a sense unique, as is shown by the theorem which follows. We make the obvious definition of order among the signed measures: $\mu_1 \leq \mu_2$ means that $\mu_1(E) \leq \mu_2(E)$ for all $E \in \mathscr{E}$.

THEOREM 6.2b. *If $\mu = \mu_1 - \mu_2$, where μ_1 and μ_2 are positive, then*

$$\mu_1 \geq \mu^+ \quad \text{and} \quad \mu_2 \geq \mu^-.$$

Proof. It follows from the relation

$$|\mu(E_r')| \leq \mu_1(E_r') + \mu_2(E_r')$$

that

$$|\mu|(E) \leq \mu_1(E) + \mu_2(E), \qquad E \in \mathscr{E},$$

whence $|\mu| \leq \mu_1 + \mu_2$. This, combined with the relation $\mu^+ - \mu^- = \mu_1 - \mu_2$, gives the required result.

The result which follows is the *Hahn decomposition theorem* for X with respect to μ.

THEOREM 6.2c. *If μ is a signed measure, there are measurable sets E, F such that*

$$E \cup F = X, \qquad E \cap F = \emptyset,$$
$$\mu^+(F) = 0, \qquad \mu^-(E) = 0.$$

Proof. There is no loss of generality in supposing that $\mu(E') > -\infty$ for all $E' \in \mathscr{E}$, and hence that

$$-\infty < \inf_{E' \in \mathscr{E}} \mu(E') = k, \quad \text{say.}$$

Let E_r be a sequence of measurable sets such that $\mu(E_r) < k + 2^{-r}$. Write

$$F_r = \bigcup_{s=r}^{\infty} E_s, \quad F = \bigcap_{r=1}^{\infty} F_r.$$

It is then immediate that

$$
\begin{aligned}
\mu^+(E_r) &< 2^{-r} & \mu^-(E_r) &> -k - 2^{-r} \\
\mu^+(F_r) &< 2^{1-r} & \mu^-(F_r) &> -k - 2^{-r} \\
\mu^+(F) &= 0 & \mu^-(F) &= -k.
\end{aligned}
$$

Writing $E = X \setminus F$, it is clear that $\mu^-(E) = 0$, since $\inf \mu(E') = k$ implies that $\mu(E') \geq 0$ for all subsets E' of E, and hence that $|\mu|(E) = \mu(E)$.

COROLLARY. If μ is a signed measure, there are disjoint sets E and F whose union is X such that $\mu(E') \geq 0$ for all measurable subsets E' of E and $\mu(F') \leq 0$ for all measurable subsets F' of F.

6.3. Absolute Continuity

If two measures μ, ν (which we take to be Borel measures, although this restriction is not necessary) are such that $|\mu|(E) = 0$ always implies $|\nu|(E) = 0$, then ν is *absolutely continuous with respect to* μ; we write $\nu \ll \mu$. If one simply says "ν is absolutely continuous," absolute continuity with respect to Lebesgue measure m is implied.

EXAMPLES. (1) The measure δ defined in Sec. 6.1 is not absolutely continuous with respect to m.

(2) Let f be nonnegative and Lebesgue integrable over every bounded measurable set E. Then the measure μ defined by

$$\mu(E) = \int_E f(x)\, dx \quad \text{if the integral exists,}$$

$$= \infty \quad \text{otherwise,}$$

is absolutely continuous with respect to m, since the integral of any function over a set of measure zero is zero. It will appear presently (Theorem 6.3d) that an absolutely continuous measure is necessarily of this form.

In the next three lemmas, it is assumed that the measures are positive.

LEMMA 6.3a. *If ν is finite, a necessary and sufficient condition that $\nu \ll \mu$ is that, given $\epsilon > 0$, there exists $\delta > 0$ such that $\mu(E) < \delta$ implies $\nu(E) < \epsilon$.*

Proof. Evidently, the condition stated is sufficient for absolute continuity. To show that it is necessary, suppose that there exists $\epsilon' > 0$ and a sequence $\{E_r\}$ of sets such that

$$\mu(E_r) < 2^{-r} \quad \text{and} \quad \nu(E_r) \geq \epsilon'.$$

Write

$$E'_r = \bigcup_{s=r}^{\infty} E_s, \qquad E' = \bigcap_{r=1}^{\infty} E'_r.$$

Then

$$\mu(E'_r) < \sum_{s=r}^{\infty} 2^{-s} = 2^{-r+1},$$

whence $\mu(E') = 0$. On the other hand, since $\nu(X)$ is finite, we have (by taking complements in the analogue of Theorem 2.3h)

$$\nu(E') = \lim \nu(E'_r);$$

and $\nu(E'_r) \geq \nu(E_r) \geq \epsilon'$ for all r.

If ν is not finite, the condition of the above lemma is not necessary for absolute continuity. For example, the measure on R obtained by writing $f(x) = x^2$ in Example (2) above does not satisfy the condition, although it is absolutely continuous. (The condition does, of course, hold in every bounded interval.)

LEMMA 6.3b. *If $\nu \ll \mu$, and ν is finite and not identically zero, there exist $\epsilon > 0$, $E \in \mathcal{E}$, such that $\mu(E) > 0$ and $\nu(E') \geq \epsilon\mu(E')$ for every measurable subset E' of E.*

Proof. Let $X = E_r \cup F_r$ be a Hahn decomposition of X (Theorem 6.2c) with respect to the (signed) measure $\nu - r^{-1}\mu$ $(r = 1, 2, \cdots)$. Write

$$E_0 = \bigcup E_r, \qquad F_0 = \bigcap F_r,$$

so that $X = E_0 \cup F_0$. Then $0 \leq \nu(F_0) \leq r^{-1}\mu(F_0)$ for all r, so that $\nu(F_0) = 0$. Since ν is not zero, $\nu(E_0) > 0$ and hence $\mu(E_0) > 0$. It follows that $\mu(E_r) > 0$ for some r. For such an r, write $E = E_r$, $\epsilon = r^{-1}$, and the requirements of the lemma are satisfied.

LEMMA 6.3c. *If ν is finite, and $\mathcal{F}$ is the class of functions which are μ integrable and such that*

$$\nu(E) \geq \int_E f \, d\mu$$

for all $E \in \mathcal{E}$, then there is a function $f_0 \in \mathcal{F}$ such that

$$\int_X f_0 \, d\mu = \sup_{f \in \mathcal{F}} \int_X f \, d\mu.$$

Proof. Let $M = \sup \int_X f \, d\mu$, and let $f_1, f_2, \cdots$ be a sequence of functions in $\mathscr{F}$ such that $M = \lim \int f_r \, d\mu$. Write $g_1 = f_1$, $g_2 = f_1 \vee f_2$ (cf. Sec. 1.5), $\cdots$, $g_r = f_1 \vee f_2 \vee \cdots \vee f_r$, $\cdots$. Then evidently $g_1 \leq g_2 \leq \cdots$, and $M = \lim \int g_r \, d\mu$. Moreover, $g_r \in \mathscr{F}$ for all r. For, let $E_1, E_2, \cdots, E_r$ be the subsets of E on which $g_r = f_1, f_2, \cdots, f_r$, respectively; then

$$\int_E g_r \, d\mu = \int_{E_1} f_1 \, d\mu + \cdots + \int_{E_r} f_r \, d\mu$$
$$\leq \nu(E_1) + \cdots + \nu(E_r)$$
$$= \nu(E).$$

Let $f_0 = \lim g_r$; then, by the generalization from m to μ of Theorem 4.1e, f_0 is μ integrable, $f_0 \in \mathscr{F}$, and

$$M = \int_X f_0 \, d\mu.$$

The result of the lemma may be refined slightly by observing that the function f_0 is essentially unique; that is, if f_0, f_0' satisfy the requirements of the lemma, then $f_0 = f_0'$ p.p. $[\mu]$. Also, f_0 may be chosen to be everywhere finite (by altering it if necessary on a set of μ measure zero).

THEOREM 6.3d (Radon-Nikodym). If $\nu \ll \mu$, there is a μ measurable function f (essentially unique and everywhere finite) such that

$$\nu(E) = \int_E f \, d\mu$$

for all $E \in \mathscr{E}$ for which $\nu(E)$ is finite.

Proof. It can easily be seen that in the real case it is sufficient to prove the theorem when μ and ν are positive and finite; for in any case X can be written (see Exercise 4) as a countable disjoint union of sets on which the measures are finite and of constant sign. The complex case is slightly less straightforward but may be deduced from the real case by a little algebra.

Let f be a function satisfying the requirements of Lemma 6.3c; then evidently the set function

$$\nu_1(E) = \nu(E) - \int_E f \, d\mu,$$

is a positive measure, and $\nu_1 \ll \mu$. If ν_1 is not zero, there exist (Lemma 6.3b) $\epsilon > 0$ and $E' \in \mathscr{E}$ such that

$$\mu(E') > 0, \qquad \nu_1(E'') > \epsilon\mu(E'')$$

for every measurable subset E'' of E'. It follows that if $g = f + \epsilon\chi_{E'}$ then

$$\nu(E) \geq \int_E g \, d\mu$$

for every $E \in \mathscr{E}$, and

$$\int_X g \, d\mu = \int_X f \, d\mu + \epsilon\mu(E').$$

This contradicts the choice of f as an element in $\mathscr{F}$ whose integral is maximal, and it follows that v_1 is zero, which is what we wanted.

COROLLARY. If $\int_E g \, dv$ exists, then so does $\int_E fg \, d\mu$, and the two integrals are equal.

Proof. The corollary follows at once from the theorem when g is a finite sum of the form $\sum c_r \chi_{E_r}$; and any integrable function can be approximated, arbitrarily closely, by such functions.

The above corollary may be regarded as, in a sense, a generalization of Theorem 5.1f. The function f now takes the place of the Jacobian J; v is Lebesgue measure m, and μ is the measure defined by

$$\mu(E) = m(g^{-1}(E))$$

(in the notation of Sec. 5.1). We have now a result which applies in very general situations, but we have, of course, no longer an explicit formula for f, as we had for J.

In general, two measures μ and v are *mutually singular*, written $\mu \perp v$ (μ is singular with respect to v, v is singular with respect to μ) if there are sets A, B such that

$$A \cup B = X, \qquad A \cap B = \varnothing,$$

$$|\mu|(E) = 0, \qquad |v|(F) = 0$$

for every measurable subset E of A and F of B. The result of Theorem 6.2c implies that, for any signed measure μ, the positive and negative variations μ^+ and μ^- are mutually singular.

LEMMA 6.3e. If μ and v are positive and such that

$$v(E) - \int_E f \, d\mu \geq 0$$

for all $E \in \mathscr{E}$ implies $f = 0$ p.p. $[\mu]$, then μ and v are mutually singular.

Proof. Consider the measure $\mu \wedge v = (\mu + v - |\mu - v|)/2$; since $\mu \geq \mu \wedge v$, it follows that $\mu \gg \mu \wedge v$. If $\mu \wedge v$ is not zero, there is a nonzero function g such that

$$(\mu \wedge v)(E) = \int_E g \, d\mu,$$

for all E for which the left side is finite, and since $v(E) \geq (\mu \wedge v)(E)$, this is a contradiction. It follows that

$$\mu + v = (\mu - v)^+ + (\mu - v)^-.$$

Let now $X = E \cup F$ be a Hahn decomposition with respect to $\mu - \nu$. Then

$$\mu(E) + \nu(E) = (\mu - \nu)^+ (E) \leq \mu(E).$$

so that $\nu(E) = 0$; and similarly

$$\mu(F) + \nu(F) = (\mu - \nu)^- (F) \leq \nu(F),$$

so that $\mu(F) = 0$.

The next theorem is the *Lebesgue decomposition theorem* for one measure with respect to another.

THEOREM 6.3f. If μ and ν are any measures, then ν can be written uniquely in the form

$$\nu = \nu_1 + \nu_2,$$

where $\nu_1 \ll \mu$ and $\nu_2 \perp \mu$.

Proof. The general case is easily reduced to that in which μ and ν are positive and ν is finite; we proceed to prove the theorem in such a case. Applying Lemma 6.3c, let f be such that

$$\nu(E) \geq \int_E f \, d\mu$$

for all $E \in \mathscr{E}$, and $\int_X f \, d\mu$ is maximal. Then if

$$\nu_2(E) = \nu(E) - \int_E f \, d\mu,$$

ν_2 is a measure which satisfies the hypotheses of Lemma 6.3e; hence $\nu_2 \perp \mu$. Writing $\nu_1(E) = \int_E f \, d\mu$, the existence of a decomposition of the type required follows at once.

If $\nu = \nu_1 + \nu_2 = \nu_1' + \nu_2'$, then $\nu_1 - \nu_1' = \nu_2 - \nu_2'$. Here the left side is absolutely continuous, and the right side is singular, with respect to μ. This clearly implies that both sides are zero, and the uniqueness follows.

The Lebesgue decomposition of a measure can be carried a stage further, in the following way: A point $x \in X$ is an *atom* for the measure μ if $\mu(x) \neq 0$. If there are no atoms for μ, then μ is *nonatomic*. It is immediate that if μ is absolutely continuous with respect to Lebesgue measure, then it is nonatomic. The converse is untrue. If μ is any Borel measure, then

$$\nu(E) = \sum_{\substack{x \in E \\ \mu(x) \neq 0}} \mu(x)$$

is easily seen to be a Borel measure also; it is the atomic part of μ. The set function $\lambda = \mu - \nu$ is evidently a nonatomic measure; and so we have the decomposition

$$\mu = \lambda + \nu$$

(easily seen to be unique) of a measure as the sum of an atomic and a nonatomic part. If the nonatomic part λ is zero, then μ is called *purely atomic*. The measure of Sec. 6.1, Example (1) is purely atomic. Applying this decomposition to the measure ν_2 of Theorem 6.3f, for the special case $\mu = m$, we have

THEOREM 6.3g. A measure ν can be written uniquely in the form

$$\nu = \nu_1 + \nu_2 + \nu_3,$$

where ν_1 is absolutely continuous, ν_2 is nonatomic but not absolutely continuous, and ν_3 is atomic.

Measures ν for which ν_2 (the singular nonatomic part) is zero are in many ways easier to handle than general Borel measures.

6.4. Measures, Functions, and Functionals

For simplicity, in this section we restrict attention to the real line R and even—for the greater part of the section—to a closed bounded subinterval $X = [a, b]$ of R.

There is a close connection between Borel measures and certain functions on X. Let μ be a positive Borel measure, and define

$$f_\mu(x) = \mu([a, x[) \qquad (*)$$

for $a \leq x \leq b$. Then it is obvious that f_μ is a finite, monotonic increasing function on X. Moreover it is left continuous ($f(x) = \lim_{h \to 0+} f(x - h)$ for all relevant x). This is an immediate consequence of the analogue of Theorem 2.3h; if $E = [a, x[$ and $E_r = [a, x - r^{-1}[$ then $E = \bigcup E_r$ and so $\mu(E) = \lim \mu(E_r)$.

Conversely, let f be a finite, monotonic increasing function. For such a function, $f(\alpha +) = \lim_{h \to 0+} f(\alpha + h)$ exists if $a \leq \alpha < b$, and $f(\alpha-) = \lim_{h \to 0+} f(\alpha - h)$ exists if $a < \alpha \leq b$. We write, conventionally, $f(a-) = f(a)$ and $f(b+) = f(b)$. Now define a set function μ ($= \mu_f$) by

$$\mu([\alpha, \beta]) = f(\beta+) - f(\alpha-)$$
$$\mu([\alpha, \beta[) = f(\beta-) - f(\alpha-) \qquad (**)$$
$$\mu(]\alpha, \beta]) = f(\beta+) - f(\alpha+)$$
$$\mu(]\alpha, \beta[) = f(\beta-) - f(\alpha+)$$

for $a \leq \alpha \leq \beta \leq b$. Then μ can clearly be defined for any finite union of intervals. The definition can now be extended to sets of the class $\mathcal{J}$, as for Lebesgue measure. The crucial step (cf. Lemma 2.2b) is to prove that if $J \in \mathcal{J}$ is written in two ways as a countable disjoint union of intervals:

$$J = \bigcup I_r = \bigcup I_s'$$

then $\Sigma \mu(I_r) = \Sigma \mu(I_s')$. It is evidently sufficient to prove

LEMMA 6.4a. If a subinterval I of X is expressed as a countable disjoint union of intervals, $I = \bigcup I_s'$, then

$$\mu(I) = \sum \mu(I_s').$$

Proof. Since

$$\mu(I) \geq \sum_{s=1}^{N} \mu(I_s')$$

for all N, it follows that

$$\mu(I) \geq \sum_{s=1}^{\infty} \mu(I_s').$$

To establish the opposite inequality, let $\epsilon > 0$ be given. Replace each interval I_s', if necessary, by a larger open interval L_s such that

$$\mu(L_s) < \mu(I_s') + 2^{-s} \epsilon,$$

and replace I, if necessary, by a smaller closed interval M such that

$$\mu(M) > \mu(I) - \epsilon.$$

The existence of suitable intervals L_s, M is guaranteed by the definition of the measure of an interval. By the Heine-Borel theorem (Theorem 1.4a), M is covered by a finite set of the intervals L_s. If Σ' denotes summation over the indices corresponding to this set, then

$$\sum\nolimits' \mu(I_s') + 2\epsilon > \sum\nolimits' (\mu(I_s') + 2^{-s} \epsilon) > \sum\nolimits' \mu(L_s) \geq \mu(M) > \mu(I) - \epsilon,$$

so that

$$\mu(I) < \sum\nolimits' \mu(I_s') + 3\epsilon \leq \sum_{s=1}^{\infty} \mu(I_s') + 3\epsilon.$$

Since ϵ is arbitrary, the required result follows.

Having defined μ for sets of the class $\mathcal{J}$, one can then define μ measurable sets just as for Lebesgue measure (Sec. 2.3), and establish that they form a σ ring. Since this σ ring includes the intervals, it includes all Borel sets, and so μ is a Borel measure.

Two distinct functions f, g may define the same measure. To take a trivial example: let $X = [-1, 1]$, let $f(x) = 0$ if $-1 \leq x \leq 0$, $= 1$ if $0 < x \leq 1$, and let $g(x) = 0$ if $-1 \leq x < 0$, $= 1$ if $0 \leq x \leq 1$. The corresponding measure in each case is easily seen to be δ, where $\delta(E) = 1$ if $0 \in E$, $= 0$ if $0 \notin E$. It is not difficult, however, to show that the correspondence between measures and functions is essentially 1-1, if suitable normalization conditions are imposed. Such a set of conditions is

$$f(a) = 0, \qquad f(x) = f(x-) \qquad (a < x \leq b).$$

These conditions are, in fact, satisfied by the function f_μ defined above.

A real or complex function f defined on X is of *bounded variation* (finite total variation) if

$$\sup \sum_{r=1}^{k} |f(x_r) - f(x_{r-1})| < \infty,$$

where $a = x_0 < x_1 < \cdots < x_k = b$ is any finite dissection of $[a, b]$ into subintervals. If this holds, then the function $|f|$, defined by

$$|f|(x) = \sup \sum_{r=1}^{k} |f(x_r) - f(x_{r-1})|,$$

where now $a = x_0 < x_1 < \cdots < x_k == x$, for $a \leq x \leq b$, is a finite positive increasing function of x. Note that, in general, $|f|(x) \neq |f(x)|$. If f is real, then the functions

$$f^{(+)} = (|f| + f)/2, \qquad f^{(-)} = (|f| - f)/2$$

are easily seen to be positive increasing functions of x also.

LEMMA 6.4b. A real function is of bounded variation if and only if it is the difference of two positive, monotonic increasing functions.

Proof. Evidently, if $f = f_1 - f_2$, where f_1 and f_2 are positive and increasing, then f is of bounded variation. Conversely, any function f of bounded variation is equal to $f^{(+)} - f^{(-)}$.

The above result is clearly analogous to the Jordan decomposition of a measure (Sec. 6.2); here also, the decomposition is not unique.

Combining the various results already obtained, we have

THEOREM 6.4c. There is a 1-1 correspondence between Borel measures on $[a, b]$ and functions of bounded variation f, normalized by

$$f(a) = 0, \qquad f(x) = f(x-) \qquad (a < x \leq b).$$

The correspondence is given by the relations (*) and (**) above.

In view of the correspondence between measures and functions, we may speak of the integral of one function with respect to another (rather than with respect to the corresponding measure), and write $\int g\, df$ for $\int g\, d\mu_f$. This is the *Lebesgue-Stieltjes integral* of g with respect to f.

The integral of one function with respect to another may, of course, be treated in a more elementary manner (Riemann-Stieltjes rather than Lebesgue-Stieltjes integrals). This approach, however, fails to provide the comprehensive convergence theorems (like those of Chap. 4), which appear in the Lebesgue-Stieltjes theory.

The situation is very similar when we consider the whole real line R. Now the correspondence is between measures and functions which are of bounded variation in every bounded interval. Such functions may again be taken to be left continuous, but now the condition $f(a) = 0$ has to be replaced by some condition such as $f(0) = 0$. If the function is of bounded variation on R, that is, if

$$\sup \sum_{r=1}^{k} |f(x_r) - f(x_{r-1})| < \infty$$

whenever $x_0 < x_1 < \cdots < x_k$, the corresponding measure is finite, and conversely. It is sometimes more convenient to normalize functions of this class by

$$\lim_{x \to -\infty} f(x) = 0,$$

rather than by $f(0) = 0$.

There are similar relations between measures and functions of bounded variation (suitably defined) in any number of dimensions.

There is another way of looking at measures which is often useful. A *linear space* is a collection of objects ("vectors") among which the operations of addition and subtraction, and multiplication by scalars (real or complex) are defined and have the usual properties. Thus, for example, the set $\mathscr{C}[a, b]$ of real continuous functions on $[a, b]$, with the usual algebraic operations, forms a real linear space. A *linear functional* is a function ϕ on the linear space to the scalar field, such that

$$\phi(\alpha f + \beta g) = \alpha\phi(f) + \beta\phi(g)$$

for all vectors f, g and scalars α, β. If an order is defined in the linear space, and $\phi(f) \geq 0$ whenever $f \geq 0$, then ϕ is *positive*. An order may be defined in $\mathscr{C}[a, b]$ by writing $f \geq g$ whenever $f(x) \geq g(x)$ for all $x \in [a, b]$.

THEOREM 6.4d (F. Riesz). There is a 1-1 correspondence between the positive Borel measures on $[a, b]$ and the positive linear functionals on $\mathscr{C}[a, b]$.

Proof. It is clear, in view of Theorem 6.1a, that every Borel measure μ determines a functional ϕ, by

$$\phi(f) = \int f\, d\mu,$$

and that this functional is positive if μ is positive.

Conversely, if ϕ is a positive functional, let F be defined by

$$F(a) = 0, \qquad F(x) = \sup \phi(f),$$

where the supremum is over all $f \in \mathscr{C}[a, b]$ such that $0 \leq f(y) \leq 1$ for all $y \in [a, b]$, and $f(y) = 0$ for $y \geq x$. Then F is a (normalized) positive monotonic increasing function, and so defines a Borel measure.

It is easy to verify that two distinct measures cannot give rise to the same functional, and that two distinct functionals cannot give rise to the same measure; hence there is a 1-1 correspondence as asserted.

The correspondence described in the above theorem may be extended in an obvious way to real measures and (with slight changes) to complex measures. The result may also be generalized from $[a, b]$ to R, and, indeed, to R^n. The appropriate linear space is now $\mathscr{K}(R)$, the continuous functions on R of compact support; there is again a 1-1 correspondence between positive Borel measures on R and positive linear functionals on $\mathscr{K}(R)$. The finite measures correspond to the bounded functionals, that is, those functionals for which there is a finite real k such that $|\phi(f)| \leq k \sup_{x \in R} |f(x)|$ for all $f \in \mathscr{K}(R)$.

Closely related to the above is the notion of *Radon measure*. A Radon measure on R is a linear functional ϕ on $\mathscr{K}(R)$ which is continuous in the following sense: Given $\epsilon > 0$ and a compact subset K of R, there exists $\delta > 0$ such that $|\phi(f)| < \epsilon$ whenever f has its support in K and $|f(x)| < \delta$ for $x \in K$. It is easy to show that every positive linear functional is continuous in this sense, so that every Borel measure is a Radon measure. However, the converse is not true. (See Exercise 19 at the end of this chapter.) The Radon measures are, in fact, the elements of the linear space generated by the Borel measures.

The definition of Radon measure can, of course, be extended to R^n or to suitably restricted (for example, open or closed) subsets of R or R^n, and, indeed, much more widely.

6.5. Norms, Fourier Transforms, Convolution Products

In this final section we mention briefly—with practically no proofs—some ideas connected with finite measures, in particular, with measures on R^n. The results generalize some already obtained in Chap. 4 for integrable functions.

Let $\mathscr{M} = \mathscr{M}(X)$ be the set of finite measures on X; as has already been remarked, $\mathscr{M}$ is a linear space (real or complex, as the case may be). It is clear that $\mathscr{L}_1 = \mathscr{L}_1(X)$ (Sec. 4.4) is in 1-1 correspondence with the absolutely continuous measures in $\mathscr{M}$; in this sense we may say that $\mathscr{L}_1$ is a subset of $\mathscr{M}$. If $\mu_{(f)}$ is the measure corresponding to $f \in L_1$, it is not hard to establish:

LEMMA 6.5a. $|\mu_{(f)}|(E) = \int_E |f(x)| \, dx$ for all $E \in \mathscr{E}$.

Consequently, if we write, for any $\mu \in \mathcal{M}$,

$$||\mu|| = |\mu|(X),$$

it follows that

$$||\mu_{(f)}|| = |\mu_{(f)}|(X) = \int_X |f| = ||f||_1.$$

It is immediate that $||\mu||$ has the properties of a norm on $\mathcal{M}$ ($||\mu|| \geq 0$; $||\mu|| = 0$ if and only if $\mu = 0$; $||c\mu|| = |c| \, ||\mu||$; $||\mu + \nu|| \leq ||\mu|| + ||\nu||$), and generalizes the norm as already defined in $\mathcal{L}_1$. Moreover, it can be shown that $\mathcal{M}$ is complete in this norm; that is, if $||\mu_r - \mu_s|| \to 0$ as $r, s \to \infty$ then there exists $\mu \in \mathcal{M}$ such that $||\mu - \mu_r|| \to 0$ as $r \to \infty$. Hence

THEOREM 6.5b. Under the norm $||\mu|| = |\mu|(X)$, $\mathcal{M}(X)$ is a complete normed linear space.

(By Theorem 4.5a, $\mathcal{L}_1(X)$ also is a complete normed linear space.)

Take now $X = R^n$. If $\mu \in \mathcal{M}(R^n)$, then any bounded measurable function is μ integrable over R^n. In particular, e^{ixy} (regarded as a function of x) is integrable, since it is continuous. (Recall that xy is to be taken as $x_1y_1 + \cdots + x_ny_n$ if $n > 1$). The function $\mathfrak{F}\mu$, defined by

$$\mathfrak{F}\mu(y) = \int e^{ixy} \, d\mu(x),$$

is the *Fourier transform* of the measure μ. It generalizes the Fourier transform of a function $f \in L_1$, as already defined (Sec. 4.6); if $\mu = \mu_{(f)}$, then

$$\mathfrak{F}\mu_{(f)}(y) = \int e^{ixy} \, d\mu_{(f)}(x) = \int e^{ixy} f(x) \, dx = \mathfrak{F}f(y),$$

by Theorem 6.3d, Corollary.

THEOREM 6.5c. The Fourier transform $\mathfrak{F}\mu$ of a measure $\mu \in \mathcal{M}$ is a bounded continuous function; and

$$||\mathfrak{F}\mu||_\infty \leq ||\mu||.$$

Proof. This is very similar to Theorem 4.6a. To begin with, observe that since μ is finite, there is a bounded set K (which may be taken to be compact) such that $|\mu|(R^n \setminus K) < \epsilon/4$; from this point onward the proof is as before.

The last part of Theorem 4.6a, that $\mathfrak{F}f(y) \to 0$ as $||y|| \to \infty$, does *not* generalize. Take, for example, the measure δ on R; it is clear that $\mathfrak{F}\delta(y) = 1$ for all y.

It may be shown that a measure $\mu \in \mathcal{M}$ is uniquely determined by its Fourier transform; that is, if $\mathfrak{F}\mu(y) = 0$ for all y, then $\mu = 0$. We do not attempt a proof here.

It may be appropriate to remark here that in probability theory the Fourier transform is usually called the *characteristic function* of the measure.

The notion of convolution product can also be generalized to $\mathcal{M}$. Suppose f, $g \in L_1$, let $\mu_{(f)}$, $\mu_{(g)}$ be the corresponding measures. If $f \circ g$ is the convolution product of f and g (Sec. 4.2) then

$$\mu_{(f \circ g)}(E) = \int \chi_E(x)\,(f \circ g)(x)\,dx$$

$$= \int (\chi_E(x) \int f(x - y)\,g(y)\,dy)\,dx$$

$$= \iint \chi_E(x + y)\,f(x)\,g(y)\,dx\,dy$$

$$= \iint \chi_E(x + y)\,d\mu_{(f)}(x)\,d\mu_{(g)}(y)$$

(where the integrations may be performed in either order). Now, this continues to make sense (formally, at least) when $\mu_{(f)}$ and $\mu_{(g)}$ are replaced by general measures μ, $\nu \in \mathcal{M}$. It can be shown that the integrals

$$\int (\int \chi_E(x + y)\,d\mu(x))\,d\nu(y), \qquad \int (\int \chi_E(x + y)\,d\nu(y))\,d\mu(x)$$

exist and are equal; and, if the resulting function of E is denoted by $\mu \circ \nu(E)$, then $\mu \circ \nu$ is a measure. This is defined to be the *convolution product* of μ and ν; it has properties very similar to those of the convolution product in L_1.

Another approach—in R, at least—is by way of the correspondence between measures and functions of bounded variation, described in Sec. 6.4. Suppose now that f and g are functions of bounded variation on R; then it is very easy to see that h, where

$$h(x) = \int f(x - y)\,dg(y),$$

is also of bounded variation. The integral certainly exists, for a monotonic increasing function, and hence any function of bounded variation, is certainly Borel measurable, and the analogue of Theorem 3.3e applies. The two definitions may be seen to be equivalent, by taking E to be the set $]-\infty, x[$ in the first definition.

In both these methods the basic properties of the product (commutativity, associativity) depend on the use of a generalized Fubini theorem (cf. Sec. 4.2). Yet another method, in which this difficulty may be avoided, is based on the correspondence between measures and linear functionals on $\mathcal{K}(R^n)$ which we outlined in Sec. 6.4. Let f and g be two positive functions in L_1, and let ϕ be the linear functional corresponding to $f \circ g$. Then, if $p \in \mathcal{K}(R^n)$, we have

$$\phi(p) = \int (p(x) \int f(x - y)\,g(y)\,dy)\,dx$$

$$= \iint p(x + y)\,f(x)\,g(y)\,dx\,dy$$

$$= \iint p(x + y)\,d\mu_{(f)}(x)\,d\mu_{(g)}(y).$$

As in the first method (but much more easily), it may be proved that this formula still makes sense, and defines a positive linear functional, if $\mu_{(f)}$ and $\mu_{(g)}$ are replaced by any positive measures μ, $\nu \in \mathcal{M}$. This may be used to define the convolution product of two positive measures, and hence of any two measures, in $\mathcal{M}$. The properties of the convolution product are rather easy to establish in this way.

It can be shown, in much the same way as for the special case of L_1, that the Fourier transform of a convolution product of measures is equal to the product (in the ordinary sense) of the Fourier transforms of the factors; that is, $\mathfrak{F}(\mu \circ \nu)(y) = \mathfrak{F}\mu(y)\,\mathfrak{F}\nu(y)$ for all y.

EXERCISES

1. If X is any set, and $\mathcal{F}$ is the σ ring of all subsets of X, verify that the following functions μ are measures on $\mathcal{F}$: (a) $\mu(\varnothing) = 0$, $\mu(E) = \infty$ if $E \neq \varnothing$; (b) $\mu(E) = 0$ or ∞ according as E is countable or uncountable; (c) $\mu(E) = n$ if E is finite and contains n points, $= \infty$ if E is infinite.

2. Show that the following are not measures (X and $\mathcal{F}$ as in Exercise 1): (a) $\mu(E) = 0$ or 1 according as E is countable or uncountable; (b) $\mu(E) = 1$ or ∞ according as E is countable or uncountable; (c) $\mu(E) = n$ if $X \setminus E$ is finite and contains n points, $= \infty$ if $X \setminus E$ is infinite.

3. Do the following define Borel measures (a) on R, (b) on $]0, 1[$, (c) on $[0, 1]$?

$$(1)\ \mu(E) = \int_E \sin x, \qquad (2)\ \mu(E) = \int_E x^{-1} \sin x,$$

$$(3)\ \mu(E) = \int_E x^{-2}, \qquad (4)\ \mu(E) = \int_E x^{-3} \sin^3 x.$$

In each case $\mu(E)$ is to be taken as ∞ if the integral does not exist.

4. Prove that the class of sets in $X \subset R^n$ which are subsets of a countable union of compact sets is a σ ring. Deduce that every Borel measure on X is σ finite. (If $X = R^n$, the proof is, of course, immediate.)

5. If $X \subset R^n$, is it true that every Borel measure on X is the restriction to X of a Borel measure on R^n?

6. In the proof of Theorem 6.1b it has been assumed that $\mu(I)$ is finite if I is a bounded interval. Justify this.

7. Show that Lebesgue measure is *not* characterized by the property of being invariant under rotations about the origin. Is it characterized by the property of being invariant under rotations about every point of R^n?

8. Prove that if $\mu = \mu_1 + i\mu_2$ is a complex measure, and $\nu = |\mu_1| + i|\mu_2|$, then $|\mu| = |\nu|$.

9. Prove that if μ and ν are real or complex measures (such that $\mu + \nu$ is defined), then $|\mu + \nu| \leq |\mu| + |\nu|$, and $|c\mu| = |c||\mu|$, where c is a real or complex scalar.

10. Show that if μ and ν are signed measures then λ_1 and λ_2, defined by

$$\lambda_1(E) = \sup(\mu(E), \nu(E)), \qquad \lambda_2(E) = \inf(\mu(E), \nu(E))$$

are not in general measures. Suggest and justify suitable definitions (direct, not in terms of the total variation) for $\mu \vee \nu$ and $\mu \wedge \nu$.

11. Show that if the ratio $|\mu|(E)/|\nu|(E)$ is bounded, then $\mu \ll \nu$, but not conversely.

12. Establish the Radon-Nikodym theorem in the complex case.

13. Show that if $\mu \perp \nu$, $\mu' \ll \mu$, $\nu' \ll \nu$, then $\mu' \perp \nu'$.

14. Prove that if the function f on $[a, b]$ corresponds to the measure μ, then $|f|$ corresponds to $|\mu|$, in both the real and the complex cases.

15. We define the function f to be *absolutely continuous* in $[a, b]$ if, given $\epsilon > 0$, there exists δ such that

$$\sum |f(x_r + h_r) - f(x_r)| < \epsilon$$

whenever $\{(x_r, x_r + h_r)\}$ is a finite disjoint class of subintervals of $[a, b]$, of total length not exceeding δ. Prove that an absolutely continuous function is (a) continuous; (b) of bounded variation.

16. Prove that a function is absolutely continuous by the definition of Exercise 15 if and only if the associated measure is absolutely continuous by the definition of Sec 6.3.

17. If g is Cantor's function, calculate $\int_0^1 x \, dg$ and $\int_0^1 x^2 \, dg$.

18. Prove that every positive linear functional on $\mathscr{K}(R)$ is continuous.

19. Prove that the functional ϕ defined on $\mathscr{K}(R)$ by $\phi(f) = \sum_{r=-\infty}^{\infty} (-1)^r f(r)$ is a Radon measure, but does not correspond to any Borel measure on R.

20. Show that if $\mathscr{M}_C$ is the set of finite complex measures on X, then

$$|||\mu||| = |\mu_1|(X) + |\mu_2|(X)$$

(where $\mu = \mu_1 + i\mu_2$) is not a norm on $\mathscr{M}_C$.

21. Prove (assuming the results stated in Sec. 6.5): (a) If μ and ν are atomic measures, so is $\mu \circ \nu$. (b) If μ is nonatomic and ν is arbitrary, then $\mu \circ \nu$ is nonatomic. (c) If μ is absolutely continuous and ν is arbitrary, then $\mu \circ \nu$ is absolutely continuous.

22. If μ_a $(a > 0)$ is the measure on R defined by

$$\mu_a(E) = \sum e^{-|r|a},$$

where the sum is over all integers $r \in E$, form the convolution product $\mu_a \circ \mu_b$. Verify that the Fourier transform of the convolution product is the product of the Fourier transforms.

Index

Index